AF692226

GIANT SILKMOTHS

To Adrien, with thanks

GIANT SILKMOTHS

COLOUR, MIMICRY & CAMOUFLAGE

PHILIP HOWSE
KIRBY WOLFE

PAPADAKIS

To the memory of my mother, who bought me my first moth book – Philip Howse

To Claude Lemaire, the indefatigable French pioneer, who painstakingly sorted and inventoried the Saturniidae, or giant silkmoths, of the Americas, publishing six large volumes and stacks of smaller publications on the hundreds of species of these beautiful moths found on these two continents. He was my friend and guide, always encouraging and helpful – Kirby Wolfe

First published in Great Britain in 2011 by
Papadakis Publisher

PAPADAKIS

An imprint of New Architecture Group Limited

Kimber Studio, Winterbourne, Berkshire, RG20 8AN, UK

Tel. +44 (0) 1635 24 88 33
info@papadakis.net
www.papadakis.net

Publishing Director: Alexandra Papadakis
Design Director: Aldo Sampieri
Editor: Sheila de Vallée
Editorial Assistant & Research: Caroline Kuhtz

ISBN 978 1 906506 25 4

Copyright © 2011 Philip Howse, Kirby Wolfe and Papadakis Publisher
All rights reserved.
Philip Howse and Kirby Wolfe hereby assert their moral rights to be identified as the authors of this work.

No part of this publication may be reproduced or transmitted in any form or by any means, electronic or mechanical, including photocopy, recording or any other information storage and retrieval system, without prior permission in writing from the Publisher.

A CIP catalogue of this book is available from the British Library

Printed and bound in China

Front cover: **African moon moth, *Argema mimosae***
Back cover: ***Agapema homogena* (female)**
Endpapers: **Indian moon moth, *Actias selene***
Half title: ***Ephiphora mythimnia* (male)**
Frontis: ***Argema mittrei* (male)**
Title page: ***Gonimbrasia krucki* caterpillar**

All images are © Kirby Wolfe with the exception of the images listed below and those in the public domain.

Front cover, endpapers, p128, p144: © Petrina Hughes; p10: © www.fossilmuseum.net, The Virtual Fossil Museum; p13: (right) © Theresa Cotter; p22, p88: © Chris Manley; p26 (bottom), p87, p94, p95, p124 (all), p125, p138, p177 (bottom): © Philip Howse; p55: © Martin Gurka; p67 (bottom): © Ghillean Prance; p76: © Tony Edgecombe; p83: courtesy of Royal Entomological Society; p86, p126: © Alexandra Papadakis; p90: © Succession Miró/ADAGP, Paris and DACS, London 2011; p91: © GardenOwls.co.uk; p97, p112 (right), p113: © Stratford-upon-Avon Butterfly Farm Ltd; p105 (right): © Jon Mercer; p106 (bottom): © Kip Loades; p112 (left): © Larry Valentine; p118: © Frank Wouters; p127: © Paul Cecil; p160: © ADAGP, Paris and DACS, London 2011; p162: © Tom D. Schultz; p164 (top): © Curt Hart; p164 (bottom left): © Roger Hall; p164 (bottom right): © Jmalik@en.wikipedia; p186: © Keren Su / Getty; p169, p175: © Michael Cook; p174: © Fastily, used under the Creative Commons license; p176 © Ch. Doumas, The Wall Paintings of Thera, Idryma Theras-Petros M. Nomikos, Athens 1992; p178: © Papadakis Publisher

We gratefully acknowledge the permission granted to use these images. Every possible attempt has been made to identify and contact copyright holders. Any errors or omissions are inadvertent and will be corrected in subsequent editions.

CONTENTS

The soul cannot think without pictures
– Aristotle

Reality identifies itself by possibilities
– René Magritte

In scientific work, those who refuse to go beyond fact rarely get as far as fact
– Thomas Henry Huxley

FOREWORD

I am glad that I was asked to write a foreword to this book as it has been a pleasure to read the informative text and to browse through the extraordinary pictures of the silkmoths. The saturniids are certainly among the largest and the most spectacular of all the moths. The pictures in this book are spectacular, but do not just look at them, for the text is also fascinating. I have often seen these magnificent *Rothschildia* moths in the field in Amazonia, but I find that there is so much that I did not know about them. However, this book is not just about moths, but about natural history in general because the silkmoths interact with so many other creatures or have devices to avoid them. The beautiful wing patterns that I have so often admired are ways of avoiding predators by mimicry, camouflage and inducing fear in potential predators. This is evolution to a perfection. Hidden in the wing patterns is a whole fauna of snakes, owls, frogs and other creatures that will stimulate fear in the predators of these moths. All of this is explained in a very readable, but scholarly text that also makes many links to history, myth and legends from around the world. The bulk of the book is about the Saturniidae family of silkmoths, the magnificent atlas, emperor and royal moths, but in the final chapter we learn so much about the true silkmoth in the Bombycidae family.

I will be looking at the silkmoths much more closely and with much greater interest the next time that I go to the tropics.

Copaxa intermediata (ma e)

pages 8/9: ***Aurivillius triramis*** **(female)**

Professor Sir Ghillean Prance FRS

COLOUR & MIMICRY

A Moth the hue of this
Haunts Candles in Brazil.
Nature's Experience would make
Our Reddest Second pale.

Nature is fond, I sometimes think,
Of Trinkets, as a Girl.

– Emily Dickinson

It is generally held that the evolution of modern families of insects began about 245 million years ago. By the early Cretaceous period, 130 million years ago, the first birds were beginning to appear: that is, feathered dinosaurs without teeth in their beaks. In that period in the Liaoning province of north-east China dust from a volcanic eruption covered muddy flood plains and imprisoned countless animals, including dinosaurs and many invertebrates, preserving their remains beautifully and forming one of the most important fossil beds in the world. Amongst the fossilised remains in the Liaoning and Hebei provinces, many of the ten thousand insect specimens found so far are recognisable as members of over a hundred families present in the world today. The predator-prey relationships between birds and insects had begun in this very distant epoch, and insects were evolving strategies to escape the attentions of the agile and keen-eyed ancestors of modern birds. Amongst the insects entombed in the oriental rocks was a large-bodied moth with a wingspan of 53 mm (over 2 in.). The size of the abdomen can easily be made out because the moth was fossilised lying on its back. It is clear that this insect would have been a good food morsel for birds and small dinosaurs. It would have been vulnerable when at rest because the size of its abdomen would have meant that it may have taken minutes for it to warm its flight muscles and become capable of flight. It would probably have been defended against predation either through resemblance to features of the plant life or the colours of the rocks and soils of its habitat, or by some other form of camouflage.

Un-named fossil moth from the early Cretaceous of Liaoning in north-east China

Perhaps, because of its colours, this moth was already difficult to see against the brown earth of prehistoric landscapes, and, in that it could have closely resembled, or even have been, an ancestor of the silkmoths belonging to the superfamily known as Bombycoidea[1] which include the common silkworm (*Bombyx mori*) and the moth *Bombyx mandarina* from central Asia that is believed to be the immediate ancestor of the silkworm.

The Romans called silkmoths *bombyx* or *bombycides*. They knew the bombyx produced silk, but were unclear about how the silk was obtained. The three most important families of the Bombycoidea for producing silk are the Saturniidae (the magnificent giant silkmoths – emperor, atlas, royal and luna moths) the Lymantridae and the Bombycidae (the latter including the common silkmoth). As we shall see, the ancient Greeks and Romans sometimes confused *Bombyx mori* silk with silk from some of the Saturniidae of Europe and the Near East. The Latin names used today were devised in the eighteenth century by the father of taxonomy, Carl Linnaeus (who also, incidentally, was the first person to claim that humans are a species of ape). He used binomial names for species, which were, eventually with many exceptions, based on ancient mythology: Roman (Latin) names for moths and Greek for butterflies.

Many of the wing patterns, colours and designs of silkmoths can be explained by mimicry of features of the environment, including those of other animals. Modern views of mimicry in animals grew from an embryonic vision in the mind of Henry Walter Bates, as a result of twelve years he spent exploring the Amazon and Rio Negro collecting the flora and fauna in the mid-nineteenth century. His insights into mimicry in butterflies greatly impressed Charles Darwin and led ultimately to numerous studies on evolutionary genetics that continue to excite biologists in the present century. He uncovered a form of deception that was widespread amongst the passion vine and long-wing butterflies (*Heliconidae* and *Ithomiinae*) of tropical America. Non-poisonous species, he realised, were mimics of poisonous species and in this way were able to profit from a degree of immunity to insectivorous birds that found the latter distasteful. Then, a contemporary, Fritz Müller concluded that there were other groupings of these butterfly species in tropical America that were all poisonous or evil-tasting and shared the same wing insignia, which then was more easily memorised, recognised, and therefore avoided, by predators. Thus Batesian mimics are sheep in wolves' clothing while Müllerian mimics are like green unripe fruit of all kinds, best avoided.

The phenomena of what are now known as Batesian and Müllerian mimicry have rarely been investigated in moths, except for day-flying species, of which there are relatively few. This is perhaps because most appear to be protected by resemblance to the backgrounds upon which they rest (i.e. by camouflage) although there is a small number of day-flying silkmoth species that are brightly coloured: the buckmoth and some of its relatives (*Hemileuca* species) share a dazzling black and white wing pattern. The colour contrast is generally greater in the males than in the females. Some species fly together at night, some by day.[2] This pattern is reflected in the caterpillars, which are black with white speckling in most species and have poisonous spines which can produce severe skin rashes. Some adult *Hemileuca* moths have prominent colour bands on the abdomen and adopt striking defensive postures, mimetic of a stinging bee or wasp. In these respects they resemble members of a quite distinct family of toxic and unpalatable day-flying moths, the tiger moths (*Arctiidae*), in some species of which flexion of a striped abdomen is accompanied by the secretion of an irritant foam from the thorax. Tiger moths, including the very poisonous white tiger moth (*Spilosoma virginica)* which has wings like white satin, and the police car[3] moth (*Gnophaela latipennis*) which has in its body poisonous pyrrolizidine alkaloids sequestered from its food plants, may therefore be examples of toxic models for some of the Hemileucas such as Griffin's sheep moth (*H. griffini)*, which might then be considered as Batesian mimics.

Alfred Russel Wallace, who conceived the idea of evolution by natural selection at the same time as Darwin, was in continual correspondence with him. Darwin was confused about the reasons for many butterflies not being as gaudily coloured as their distasteful caterpillars and asked Wallace's opinion. Having been informed by a correspondent that a white ermine (tiger) moth, *Spilosoma lubricipeda*, was avoided and never eaten by aviary birds or turkeys, Wallace realised that a white moth at dusk is as conspicuous as a coloured caterpillar in daylight, and immediately wrote to Darwin about it. Darwin replied, "I never heard anything more ingenious than your suggestion.... It warms one's very blood to see a theory thus almost proved to be true."[4]

below left: **Male Griffith's sheep moth, *Hemileuca griffini*, – a moth palatable to most birds**

below right: **Police car moth (*Gnophaela latipennis*) from North America, a possible toxic model for Hemileuca moths**

A difficulty of having these two separate categories of mimicry is that one grades into the other. One insectivore's meat is another's poison and it is this that creates an ecological balance. To illustrate this point, it is well known that the famous monarch butterfly (*Danaus plexippus*) is toxic to predators because the caterpillar sequesters in its body cardiac glycosides from its food plants, which cause a variety of unpleasant symptoms when ingested. The food plants are milkweeds, which are generally toxic to all animals, but the monarch has evolved an immunity to them and thereby secures for itself an almost exclusive food source and at the same time protection from predation.

It was generally assumed that the viceroy butterfly (*Limenitis archippus*), which does not feed on milkweeds but closely resembles the monarch in colour and patterning, is a Batesian mimic – until, that is, it was discovered that it, too, is toxic. To muddy the waters still further, it was also found that not all monarchs are equally toxic – different species of milkweed contain differing amounts of glycosides with the result that some monarch populations are actually palatable to birds. Batesian and Müllerian mimicry therefore seem to be extreme aspects of the same phenomenon and the evidence for either is only ever circumstantial. They both depend upon the predators, usually birds, learning the hard way that insects with certain shapes and colour patterns are distasteful and poisonous to some degree. Several studies have shown that birds can remember the association for many months after a single shock encounter.

Bright contrasting colours are generally found in butterflies and moths that are poisonous or produce toxic secretions. They are not always black and white, like many hemileucine moths, but include other colours. Red coloration of the hind-wings is common in many insects apart from Lepidoptera (butterflies and moths), including grasshoppers, bush crickets, mantids, and phasmids (stick and leaf insects). Bright coloration with a presumed warning function is known as *aposematism* or (especially when colours are suddenly revealed) "flash coloration". The effect of brusquely exposing bright wing colours is to produce a startling and radical change in the visual image, such as we may experience when a bright light is switched on at night. Flash coloration is effective mainly in day-active insects. In very low light conditions the colour-sensitive cones of the vertebrate eye become unresponsive and hence, to cope with the situation, we find that bright circular ring patterns forming easily detectable "eye-spots" are commonly encountered among saturniid moths.

It is not only a model with a toxin inside it that may give rise to the evolution of mimetic forms of other species, but a model which is defended by bright colours and behaviour. *Hemileuca electra* and the subspecies *mojavensis* are good examples of non-toxic moths with warning colours, in which the maximum colour contrast is achieved by black-white

pages 14/15: ***Hemileuca electra mojavensis*** **(male); a North American day-flying male with warning coloration**

opposite: ***Eochroa trimeni*** **(male) from southern Africa; a day-flying toxic moth with warning coloration**

fore-wings and a red-black or red-yellow pattern on the hind wings. They illustrate the point that these moths are non-toxic and therefore appear to be defended by colour contrast and reflectivity of light alone.

Agapema silkmoths of North America have black (or grey) and white wing patterns similar to the *Hemileuca* moths. Although the male moths are conspicuously patterned, there is no evidence that they are poisonous. It is possible, however, that the high colour contrast is in itself a deterrent to opportunistic predators because of the startle effect. *Agapema* moths are nocturnal, as are most white and black Hemileucas, but the female Agapemas are predominantly grey in colour, making them inconspicuous.

Some aposematic silkmoths flex their conspicuously banded abdomen in a motion that calls forth images of the stinging movements of bees and wasps. They usually fall to the ground and stay immobile for a time, maintaining the stinging posture.

This is not, to us, on first impressions, a very convincing form of mimicry – but nature did not intend it to be so because it is one that exploits visual ambiguity. This has been termed "Satyric mimicry".[5] It is common among hoverflies which include many mimics of bees and wasps. There are, for example, some near perfect hoverfly mimics of wasps and yellow-jackets (judged by the human eye) but others which appear to be very poor copies – and despite being so it has been shown in experiments that they are treated by birds in the same way as good mimics. The explanation appears to be that the ambiguity distracts predators for a few milliseconds, long enough in nature to allow the insects to escape, or even discourages the predator from further inspection. It then follows that the model for this mimicry need not be toxic, as in the similar scenario of a moth that appears to be an imperfect leaf mimic (of which there are many examples). There is some evidence, though, that the aversion to yellow and black wasp stripes is innate, at least in some birds, and does not have to be acquired by experience when they are young.

Agapema galbina (male and larger female), Mexico

There are obvious benefits in terms of survival for a moth that mimics another insect capable of inflicting a painful sting, but there are other examples of moths that are mimics of insects other than other, possibly poisonous, moths. We see this among the *Citheronia* moths, in which some species closely resemble cicadas, with a large patterned thorax and the typically prominent wing veins of the cicada marked in bright orange or yellow.

No evidence has been recorded that any cicadas are toxic;[6] their defence consists of loud squawks when they are trapped. In some cases the sound intensity (approximately 110dB) reaches the human pain threshold, which has been shown to give them limited protection from grasshopper mice.[7] The long proboscis is designed for penetrating the bark of trees, and can be a hazard if they are not handled carefully, but the benefits of the cicada-like Citheronias may simply lie in the ambiguity of the image they present to the observer.

above left: ***Citheronia vogleri*** **(male)**

above right: ***Hemileuca electra electra*** **(female). Two moths in which the banded abdomen is flexed in a mock stinging display**

opposite: ***Citheronia azteca*** **(male), a presumed cicada mimic**

The concept of satyric mimicry has been extended to a form that can be called "masking" or "prosopal" mimicry.[8] "Mimicry" has the same Greek root, *mimesis*, as "mime" and "imitation". To Aristotle mimesis meant a simulation of nature, shown in art and acting, in which the actor would take on part of the essence of another being – an effortless transformation for great actors. Actors in ancient Greece and Rome

would often play several different roles, and the transition between personalities was commonly achieved with the assistance of masks. The mask-makers of these early civilisations had the same skills as cartoonists today, selecting and magnifying the peculiarities of facial expressions that created characters that were demonic, jovial, aristocratic, aged, drunkards, beggars, etc. The mask was known as a *prosopon* (Greek) or *persona* (Latin).

We are often deceived in what we see by shadows, the play of light on a surface, and what looks like a familiar face. The neurophysiologist Richard Gregory hypothesised that the human brain sets up models of objects, analogous to the 3-D virtual image that can be generated on a computer and viewed from any angle. The brain then matches a visual image with the immeasurably large library of such templates and finds the best fit. If the information reaching the visual cortex is inadequate or contradictory in some way the brain may make a "best guess", as a computer does when you feed

The death's head hawk moth, *Acherontia atropos*. The skull pattern on the thorax is an example of pareidolia – seen from the perspective of a bird it is an illusory image of a giant hornet

pages 24/25: Edward's Atlas moth, *Archaeoattacus edwardsii*, showing features of snakes, claws, etc., embedded in the wing pattern

the search engine a misspelt key word and get it wrong, in which case you may see a cat on the sofa instead of a black scarf, or the Virgin Mary in a piece of toast, (or, in a notorious case, Mother Teresa's face in a bun).

This tendency to see things that we like to think are there is known as *pareidolia*. One example of this attaches to the Japanese Heike crab, which many believe has the face of a Samurai warrior on its carapace. According to an ancient legend, defeated warriors plunged into the sea to atone for their disgrace and were turned into crabs. Biologists such as Julian Huxley argued that this accelerated the survival and evolution of the crabs because superstitious fishermen threw back all the Heike crabs they caught. However, this pattern occurs on a variety of other species of crab and simply reflects the positions of the anchor points of the internal muscles (remember: all arthropods, including crustaceans, have a skeleton that is external, not internal, and the muscles are internal to that). This then is a case of pareidolia, in which people see an approximation to an image they have been primed to see and generalise from that.

The death's head hawk moth (*Acherontia atropos*), made famous by the book and film *The Silence of the Lambs* represents another example of pareidolia. The marking interpreted by many as a skull and cross-bones is nothing of the sort, and most likely is an inverted image of the head of a giant hornet, which is easily recognisable when it is viewed from the head end.[9]

The Atlas moth (*Attacus atlas*, see Chapter 7) is a somewhat different case. Here, it is easy to see the image of a snake on the borders of the wings. When the insect is molested it falls to the ground and flaps its wings slowly moving the tips in a figure of eight path and the "snake" appears to writhe about. There is no explanation for this in anatomical terms and the simplest explanation (in the absence of experimental proof) is that this is part of a defence strategy that is a deterrent against predators such as birds and monkeys. It is not an image of photographic quality – more of a cartoon image with a limited number of abstracted features.

In addition to the phenomena that Bates and Müller described, the wing patterns and colours of nocturnal moths are determined by other factors dictated by natural selection. Alongside the obvious camouflage designs, which we will examine in the next chapter, it has now been found[10] that both butterflies and moths carry mimetic patterns of other animals on their wings, particularly birds, snakes and other reptiles. This has been overlooked in the past because only the key features of such predators: eyes, teeth, beaks, feathers, scales, etc., are embedded in the overall wing patterns. These, we rarely notice because we have Gestalt vision that enables us to recognise objects without examining their details first, and also because we are accustomed to seeing images of butterflies and moths set as museum specimens in unnatural positions that disrupt the designs displayed by the live insects.

When we look at butterflies and moths we automatically assume that birds see what we see. This is a huge assumption to make: we know more about the moon than we know about what goes on inside a bird's brain. The bird brain has evolved along a different path from the human brain; it is anatomically different and while it has a cerebrum does not have the huge enveloping corrugated cerebral hemispheres characteristic of the brain of higher mammals. The bird's eye has four and sometimes five different types of colour-sensitive cones compared with our three. It also has double cones and oil droplets of various colours in the cones, the function of which is not clear.

Animals do not experience the same sensory world as we humans. Some people are colour blind, we are all insensitive to ultraviolet, and it is impossible to know whether the sensations of colour that we experience are the same as those of the next person. Some people are affected by synaesthesia: they see colours associated with musical chords, letters of the alphabet are seen in different colours, colour may be coupled with the hearing of sounds and taste with touch, etc. The generally accepted theory currently is that babies are synaesthetic and part of their development involves coupling the input from their sense organs with appropriate sensations. We do not know, and perhaps never

will know, whether birds are synaesthetic, and birds are the most important predators of butterflies and moths, apart from bats (which do not depend on vision). However, it is known that the magnetic sense organs that migratory birds use to detect the earth's magnetic field send their information to the visual part of the brain, and that birds see the magnetic field as patterns superimposed on the visual field.[11] Experimental evidence from research on primates suggests that their visual system is similar to ours, but there is little information on small mammalian insectivores.

It is clear, then, that because of the differences in modes of perception, visual features of moths' wings that may not be immediately apparent to us may stand out sharply to a bird. It is often found that the defensive display of a moth is not purely visual but is accompanied by rocking or wing flapping and sometimes by the secretion of evil smelling liquids or clicking sounds. These non-visual aspects of the performance may add to or, conceivably, magnify the visual effects. At all times, we have to bear in mind that the defensive or intimidating displays of moths have been designed by the operation of natural selection to be clearly perceived by their principal predators, which are birds and bats, not human beings.

A great deal of what we do know about visual perception in birds comes from observations and experiments with models carried out by the founders of ethology, the scientific study of animal behaviour, Konrad Lorenz and Niko Tinbergen. Every biology student will have come across Tinbergen's study of the feeding response of the herring gull (*Larus argentatus*). The gull chick pecks at the red spot at the base of the lower bill of the parent, and then obtains a meal of regurgitated food. Using card models of various shapes, colours and configurations, Tinbergen was able to show that the chicks would peck to any red mark on a light-coloured background. They behaved as if they did not see the parent at all – their visual world was constrained to a small detail. As young birds mature they develop the ability to recognise an increasing number of visual signs, but all the evidence suggests that they respond first to detail and not, like us, to the totality of the image of another animal or object.

We find it difficult to read anything that is upside down. A useful technique for clearing a train compartment is to read a newspaper upside down. The surrealist artist Salvador Dalí produced a wonderful painting of swans on a pool. He painted in the reflections, and the neck of a swan became the trunk of an elephant, and the sail-like wings became its broad ears. In this way Dalí shows us that our world is one in which the position of an object with respect to the vertical axis, with "up" being in a vertical direction counter to the pull of gravity, is of great importance in enabling us to recognise that object. This orientational aspect is very evident in museum specimens or identification guides. A moth is pinned in a cabinet with the wings set in a horizontal plane and the head up

Graffiti in a park. The anonymous artist has figured some of the main threat symbols that are incorporated into the wings of saturniid moths: sharp teeth, claws, eyes and a reptilian profile

opposite: ***Caligula simla*** (male) from east Asia showing the face of an owl or eagle with open beak and raised wings, not evident from the "normal" perspective

or facing directly away from you. If all species are set like this it makes it easy for taxonomists to compare them, and extending the wings forwards in the horizontal plane makes it possible for them to focus on all the minute features of pattern, venation, hairs, etc. But what you see is remote from the way the moth looks in life, especially to a bird that may see it "upside down".

The different angles at which birds approach potential prey is illustrated by observations made on the peppered moth (*Biston betularia*). One of the most frequently used examples to illustrate natural selection is that of the apparent adaptation of this moth to a changing environment. The moth, which has whitish-grey wings peppered with black dots is well camouflaged resting on bark with lichen but early in the nineteenth century there was an extremely rare melanic (almost completely black) form recorded. H.B. Kettlewell found that in the industrial Midlands of England at the end of the nineteenth century selection pressure had led to over 90% of the moths existing as this melanic form, which was so well camouflaged as to be almost undetectable against tree bark blackened by industrial smoke pollution.

When Kettlewell's work was published his evidence was rejected by some critics. Niko Tinbergen then collaborated with Kettlewell to obtain filmed evidence of the role that birds play in changing the moth populations. In rural Dorset birds mainly select melanic moths and in the Midlands they go for normal moths.[12] It was not difficult to obtain this evidence, and in the process the biologists made some interesting observations on the foraging techniques of birds. Flycatchers and yellowhammers hovered up and down the tree trunks close to the bark, but failed to find the typical moths on unpolluted trees. Song thrushes looked up from the ground, robins and great tits inspected from a vantage point and then flew towards the moth and made a detailed search at close range, and nuthatches ran up and down the trunk and were able to detect moths by their profile close to the bark. Clearly, birds may encounter moths from any angle and have the ability to identify them from any particular viewpoint.

In Polythysana, the brighter males fly during the day and the plainer females after dark. We visited Chile in March during their early Fall, when this genus (three species) flies. In the south central part of the country we stopped at noon to stretch our legs, and I immediately noticed the males swooping around over the low treetops. I assembled my butterfly net with tall extension poles, clumsily trying to guide it towards the darting moths when a male suddenly flew into it! It was P. cinerascens. As impossible as it had seemed, if I had not tried I would definitely not have caught it. So I decided to return that night to a small pasture by the road where we had caught the male, and set up my generator and mercury vapour light on a sheet spread on the ground. Within minutes a fat female of P. rubrescens landed squarely on the sheet. It was a very happy outcome.

– Kirby Wolfe

In summary, insectivorous birds are the most important visually hunting predators of butterflies and moths in most ecosystems. Their visual perception differs from our own: they can see into the ultraviolet range. They are generally short-sighted and mostly rely initially on monocular vision for detection of prey so that they are liable to confuse the size of an object and its distance. Their eyes are only a few inches above the level at which they are perched so they see objects from a different perspective from us, and very often from a different angle. They focus on detail, at least initially, and do not see the "whole picture". Most importantly, in taking rapid decisions about what is potentially edible and what is not they may be momentarily confused by ambiguities in what they see.

Moths, unlike butterflies, do not, on the whole, benefit from bright colours unless they are day flying. Nocturnal birds and other vertebrates have few colour-sensitive cones in their retinas; instead these are packed with rods which have high sensitivity in low illumination. As light dims, the cones cease to respond to red first, so colours at the red end of the spectrum are not easily seen at dawn and dusk. Generally, attraction of the sexes, courtship and mating are controlled by chemical signals (pheromones) and vision plays little or no part.

The habit of many moth species is to hide during the day and remain immobile, usually well camouflaged, unless they detect a threat. If that happens, large-bodied moths in particular are at risk because, like geese and swans, it takes them time to become airborne. They need to generate power for lift-off, but they are cold-blooded and must first raise their body temperature to over 30°C before their muscles can begin to function effectively. The robin moth (*Hyalophora cecropia*) for example may take up to eight minutes to raise its body temperature from 15°C to over 30°C at which level it can begin to fly. The warm-up process is called "shivering" and that is how it appears – the wing muscles contract rhythmically and almost isotonically (without changing length) so that energy is converted into heat rather than movement.

During warm-up, moths are obviously highly vulnerable to predation. Trapped vertebrates can threaten with teeth, beaks and claws, but moths have no such weapons. Instead they have images of these and of other predator features painted like graffiti onto their wings. The effectiveness of these images as threats is often accentuated and made more realistic by rocking movements or by sliding the wings backwards and forwards on the horizontal axis. In our world, graffiti can be very intimidating. Some with images of men with guns inscribed on school walls in the USA have caused the schools to be shut down, and we are all familiar with the power of posters for Dracula and vampire films.

Polythysana cinerascens (male), from Chile. The fore-wings have images of finch heads in profile

pages 30/31: *Polythysana rubrescens* (male) in which the face of a small mammal can be clearly seen. Notice the white legs, which with half-closed eyes become canine teeth. At the same time the moth resembles the head of a king vulture with the same colour pattern and the body outlining the forward-facing beak with open mouth behind

The recognition of the familiar in newborn or newly-hatched animals depends to a large degree on the learning process known as "imprinting", often shown by a following response, of which there have been many studies in ground-nesting birds such as ducks and geese. It is vital for them also, to be able to recognise their parents as soon as possible after hatching and distinguish them from other beasts and objects they see. This means that all birds develop with an imprinted image of their species characteristics, basically those represented on the head of the parent.

Deception and momentary confusion may therefore arise in a bird that sees the image of a bird's head or wings embedded in the wings of an insect, and we will see many examples of these images on the wings of saturniid moths. In reading to this point you may not have noticed that the head and beak of a bird is engraved on each fore-wing of *Hemileuca electra* shown on page 20. Equally striking examples seem to leap from the page on the fore-wings of the three Chilean species of *Polythysana*. In these moths the brighter coloured males fly during the day, and the females at night.

The males of *Polythysana cinerascens* and *P. rubrescens* are fearsome-looking creatures. The two bird heads with finch-like beaks slide apart to reveal the mask of a red-eyed rodent, visible more clearly on inversion, so from whichever aspect you view the moth the illusion is momentarily startling.

We will see other examples of embedded bird images in the subsequent chapters, and also examples of mimicry of claws, teeth and fur of mammalian predators. *Copaxa sapatoza*, for example, has a plethora of long brown hairs extending over the wings from which emerge, as if painted by a skilled artist, four vicious-looking feline claws. The "claws" are "windows" in the wings, lacking scales, but light-reflecting at certain angles. These are well-developed in *Rothschildia* moths (Chapter 7).

In certain species, claw and hair features are combined with eye images, again creating an intimidating image of a mammalian head – in *Copaxa intermediata* the nose and mouth of a small mammal are clearly recognisable.

Various species of *Saturnia* and *Agapema* from northern America have the white and black (or grey) patterns on their wings which suggest that they may be part of a mimicry ring that includes Hemileucas. They are not known to be poisonous, but they do, however, present the colour patterns found on the North American white-tailed deer (*Odocoileus virginianus*). This once very common deer with a widespread distribution, has a grey coat in the winter and a brown coat in the summer, and its most conspicuous feature is a white band of fur that envelopes each eye.[13] This feature we see in some *Hemileuca* moths, but it is reproduced with more startling effect in *Saturnia albofasciata*, in which the white band around the nostrils of the deer is also hinted at. In making these comparisons we have to take into account the difficulty already mentioned that birds have in distinguishing between large objects at a distance and small objects close to. We would never mistake a large moth for a more distant deer, but we do not have the same visual processing mechanisms as birds or reptiles.

Dysdaemonia fosteri (male), a species in which the face mask of a small mammal (with ears and whiskers) becomes visible when viewed in the head-down position

opposite: ***Copaxa sapatoza*** (male) – a giant feline foot and protruded claws?

The examples we have been looking at so far have been from the New World, but in the Old World there are parallel phenomena. *Aurivillius triramis* and *Gonimbrasia krucki* from Kenya have the colours, the hair coats, the eyes and the facial lines that are commonly found in members of the cat family. Hugh Cott[14] wrote that at close range such resemblances "must afford the most terrific sight to birds and other insect-eaters that are themselves the favourite food of such carnivores as genet, serval or civet".

Another example of vertebrate mimicry, again from Africa, is that of *Bunaea alcinoe*, which presents a mask with a copy of the pronounced and fairly straight eyebrow ridge typical of that of a drill or baboon and eyes, quite unlike owl eyes, to match.

Eacles barnesi seems on first sight to have two small eye-spots on the thorax, but when seen in the head-down position it is possible to see a sudden metamorphosis in which these become nostrils and the head of a mammal, much like a bear, becomes visible.

Extensive mimicry also exists among the caterpillars of the saturniids. One of the most remarkable examples is the larva of *Bombyx mandarina*, the ancestor of the common silkworm, in which the inflated head appears as that of a snake and the head viewed from the side resembles that of a frog.

above: ***Saturnia mendocino***, wing-tip of a female from California

top: ***Saturnia albofasciata*** (male with larger female)

opposite top left: ***Hemileuca stonei*** (male) from Arizona

opposite centre left: ***Eacles barnesi*** (male) from South America

opposite top and centre right: Caterpillar of ***Bombyx mandarina***, presenting itself as the head of a snake from the front *(top)* and as a small frog from the sideview *(centre)*

opposite bottom: ***Gonimbrasia krucki*** from Kenya. Two views of a male with wings bearing facial images of members of the cat family

Bunaea alcinoe (male), from Africa with 'eyes' and 'eyebrow ridges' resembling those of a drill or baboon

CAMOUFLAGE & CONCEALMENT

The mysteries of mimicry had a special attraction for me. Its phenomena showed an artistic perfection usually associated with man-wrought things … I discovered in nature the nonutilitarian delights that I sought in art. Both were a form of magic, both were a game of intricate enchantment and deception. – Vladimir Nabokov

Liu Bolin, a Chinese artist known as the "Invisible Man" has astonished millions of people with his work. He uses body painting as camouflage. His technique is to remain absolutely still against backgrounds such as supermarket shelves, theatre seats, city landscapes, etc., while two assistants paint him in colours and patterns that make him part of the background like the missing piece of a jigsaw puzzle. Only when he moves do passers-by detect his presence. It is remarkable, if not ironic, that he has achieved worldwide fame when natural selection has rendered innumerable species of moths and other animals similarly "invisible".

There are close parallels between the types of uniforms developed for military use and the disguises adapted by insects for warning and defence. The British army (and the armies of certain other nations) in the seventeenth to nineteenth centuries wore uniforms that were bright red. Initially these were dyed with madder and later with cochineal – a bright red dye extracted from the cochineal insect. The red coats were

Syssphinx albolineata (female), Texas, north-east Mexico. A moth showing camouflage by colour, disruptive patterning, with the alternative defence of warning colours that can be brought into play by exposing the hind wings

worn with contrasting colours, such as white or black. One purpose of this style of uniform was to advertise the presence and threat of the formidable British soldiery, and another was to make the soldiers visible in what became known as the "fog of war". The latter was due in part to the use of gunpowder that produced a lot of smoke and made it more likely that troops would attack their own comrades if they could not be easily identified.

Smokeless gunpowder came into use in 1880 during British rule in India and the red uniform was dropped because soldiers wearing it could then be more easily targeted, particularly with the longer range guns then developed. Ultimately it was phased out for general use because of the high cost of the cochineal dye. The uniform in India was replaced by khaki cloth, the colours of which blended into the environment, particularly when the dye began to run when wet, as Rudyard Kipling described in his short story *The Arrest of Lieutenant Golightly*: "The khaki colour ran too – it was really shockingly bad dye – and sections of Golightly were brown, and patches were violet, and contours were ochre, and streaks were ruddy red, and blotches were nearly white, according to the nature and peculiarities of the dye. When he took out his handkerchief to wipe his face and the green of the hat-lining and the purple stuff that had soaked through on to his neck from the tie became thoroughly mixed, the effect was amazing."

Thus it was the khaki cloth, above all, that gave rise to camouflage battledress and disguise of military vehicles and installations by use of colour and random patterning. The camouflage works by concealing the outlines of people and objects, and breaking up solid form into a kind of visual chaos, a phenomenon that has been explored in the works of many artists, including Andy Warhol.

Insects, including bugs, mantids, stick insects, butterflies and moths, mimic every form of natural vegetation and natural substratum: green leaves, dead leaves, bark, thorns, twigs, grasses, fungi, mosses, lichens, seeds and flowers, stones, bare earth, etc. It is difficult, when discussing mimicry, to avoid the semantic trap of choosing a term that is not teleological – suggesting that nature has a purpose, which then leads to assumptions about the reasons for things. There has been much argument amongst biologists about whether to use the terms mimicry, protective resemblance or camouflage, but ultimately the neologism *crypsis* has prevailed. Nevertheless, the principles at work in military camouflage are basically the same as those operating in animals which are, well, camouflaged (or disguised – as in the use of make-up for actors) and which is the meaning of the antecedent French word *camoufler*).

Dead leaves change colour as the chlorophyll is gradually degraded and withdrawn from the leaf, leaving the orange, red and yellow pigments exposed. William Hamilton, who

Dead leaf mim cry in *Lonomia achelous* from Brazil, a male and larger female

made many valuable contributions to evolutionary theory, controversially proposed that one function of the colours of autumn leaves in temperate forests, epitomised by the fall colours in New England, was to deter insect attack. He regarded the colour change as an advertisement for the fact that the leaves no longer contain a rich supply of nutrients for herbivores. It has also been suggested that many herbivorous insects, which are camouflaged resembling green leaves, then become visible to predators.

Leaf fall in tropical forests happens mainly during the dry season, although some trees shed their leaves at irregular intervals throughout the year – a strategy that can make existence very difficult for insects that feed on them. Dead leaves are the staple diet of the forest, which feeds itself with a variety of assistants. Leaves are digested by micro-organisms, mainly bacteria and fungi, and by small invertebrates among which termites are of great importance. The termites themselves harbour micro-organisms in their guts that can break down the cellulose cell walls of plant tissues. Many of the mound-building termites of the tropics and the leaf-cutting ants of the New World construct fungus gardens in their nests. The symbiotic fungi, which are kept in pure culture by the insects, convert the plant material assembled for them into protein in their own tissues, which the insects then feed upon. This actually mirrors the commercial production of mushrooms.

Not surprisingly, we find that many insects have a form that mimics dead leaves; dead leaves with typical leaf shape, colour and vein patterns, with holes eaten out of them, with fungal growths on them, and dead leaves that appear dry and curled, or hang in bunches.

Let us start with *Lonomia,* which embodies several camouflage techniques. Hold the book away from you and half close your eyes and you see two dead, leathery leaves with thick mid-ribs. The axis of symmetry of the moth has been changed by 90°, helped by the head and abdomen being subtly obscured. The wings overlap in such a way as to conceal the body completely, and the outline of the head is confluent with the wing outlines. Disguise of the vulnerable head and body is completed by the faint patterning, which suggests a fungus growing over decaying wood. Although the adult moth is protected by disguise alone, contact with its caterpillar can be fatal, as we shall see further on.

Cricula andrei and *Copaxa litensis* have gone a step further. A series of apparent holes in the leaf pattern simulate those made by the feeding of termites, springtails, beetles or such creatures. These holes are optical illusions; in fact pigment scales are lacking at this point on the membranous wing surface, making it transparent.

opposite top: ***Copaxa litensis*** **(female)**

opposite bottom: ***Cricula andrei*** **(female) against a dead leaf background**

pages 44/45: ***Copaxa simson*** **(female)**

Contemplation of the high degree of perfection of mimicry shown by these moths leads us into a new world. If natural selection can achieve such copies of the leaves among which moths rest, it should not be surprising to find copies of other features of the environment, such as we have seen in the previous chapter. An initial search among a wide range of butterflies and moths[1] has shown that there appear to be no limits to the masquerades of these creatures.

Evolution appears to have followed two paths from here: coalescence of the feeding holes into a circular window that starts to resemble an eye. Or the addition of markings that mimic the action of moulds on the leaf surface, coupled with colour patches that suggest dried out patches on the leaves as in *Eacles penelope* (p.55), in which the leaf holes are even replicated over the body, camouflaging it still further. At the same time, basic features of camouflage come into play: the outline of the moth's wings is obscured and disrupted by light grey patches and the black line as well as the line of spots, which changes the apparent axis of symmetry and helps to conceal the body of the moth itself.

Another form of camouflage that was developed for military use, especially in the disguise of battleships is known as "dazzle painting". Bright zebra stripes, usually in zig-zag patterns, are painted running across the sections of the ship. The eye tends to follow the meanderings of these markings, which effectively obliterate the outlines of the ship and its main structures.

An effect similar to dazzle is achieved in some moths by black lines on a pale background, as in *Copaxa simson*. A child with a felt pen seems to have been let loose on *Copaxa simson*. Black lines over the wings set a conundrum: do they mark the edges of leaves, midribs, a long-legged spider, fungal growth? Does it matter? What they do *not* do is delineate the form of a large-bodied moth.

Border-lines accentuated by contrasting colour bands on the wings, as in dazzle painting, draw attention away from the wing outline. This can highlight the outline of another image on the wings and the graded density of pigment, like countershading on a picture, can create the illusion of three-dimensionality, as in the African species *Pseudobunaea irius*, and *Epiphora* species. The addition of brightly coloured or reflective spots in the centre of the wings helps further in directing the gaze to the imposed image. Those spots, sometimes coloured, sometimes black, and in the case of Epiphoras highly reflective, presage the evolution of large eye-spots. The intimidating effect of the eye-spots of many moths is magnified when the insect displays by inclining its wings forward.

opposite top: ***Lobobunaea angasana*** **(female), an African species with a dense black "pupil" in the eye-spot**

opposite bottom: ***Epiphora mythimnia*** **(female), from South Africa**

Lichen-like ornamentation of a *Citheronia lobesis* caterpillar

opposite top: ***Syssphinx quadrilineata*** **caterpillar. A line marks the edge of an apparent spiny cactus leaf**

opposite bottom: ***Citheronia vogleri*** **(female)**

The camouflage technology (if one may call it that) of caterpillars is extraordinarily varied. Many means have evolved of disrupting the typical body outline; these include border-lines, which may be straight, undulating, or broken into repeated segments. This makes it appear that the body is separated into two or more parts, each of which can be interpreted in different ways. The caterpillar of *Citheronia lobesis* has a sharply defined, brightly coloured dorsal saddle zone resembling an encrusting lichen, contrasting with the dark-coloured legs and ventral surface that is almost invisible against the stem on which it rests.

More commonly, "dazzle" camouflage is achieved by rows of bright spots, as in some Citheronias.[2] The black and white pattern may have a double function here: a high-contrast warning signal that also disrupts the body outline.

Concealment through the use of irregularly shaped patches of colour is used extensively today in military uniforms and on tanks and armaments, but is anticipated in moths such as *Dirphia avia* and the caterpillar of *Titaea lemoulti.*

right: ***Dirphia avia*** **(male) – disruption of outline, but perhaps, as the Latin name implies, there are silhouettes of bird heads here**

below: ***Titaea lemoulti*** **caterpillar**

opposite: ***Micragone herilla*** **(male). A superb bark-mimic, but with an embedded scorpion image**

The evolution of a seemingly random pattern of markings on a dull background, such as is frequently used on military camouflage, can be seen in some moths, such as *Citheronia maureillei* and *Micragone herilla*, but moth evolution has not overcome the constraints of bilateral symmetry and the symmetry makes it possible for us, at least, to see a pattern (as it does in Warhol's prints). Not only that, but the *Micragone* carries the image of a scorpion.[3]

The larger butterflies and moths tend to overlap in size with small birds. With a slight change in the patterning, the yellow highlights that are seen on the wings of *Citheronia maureillei* coalesce into the form of feathers in its cousin *C. laocoon*, and one sees the wings of a yellow bird.

In one remarkable form of mimicry that is found particularly amongst members of the *Copaxa* family the patches of pale colour simulating dried-out parts of the "leaf" also create a new image on the wing canvas:

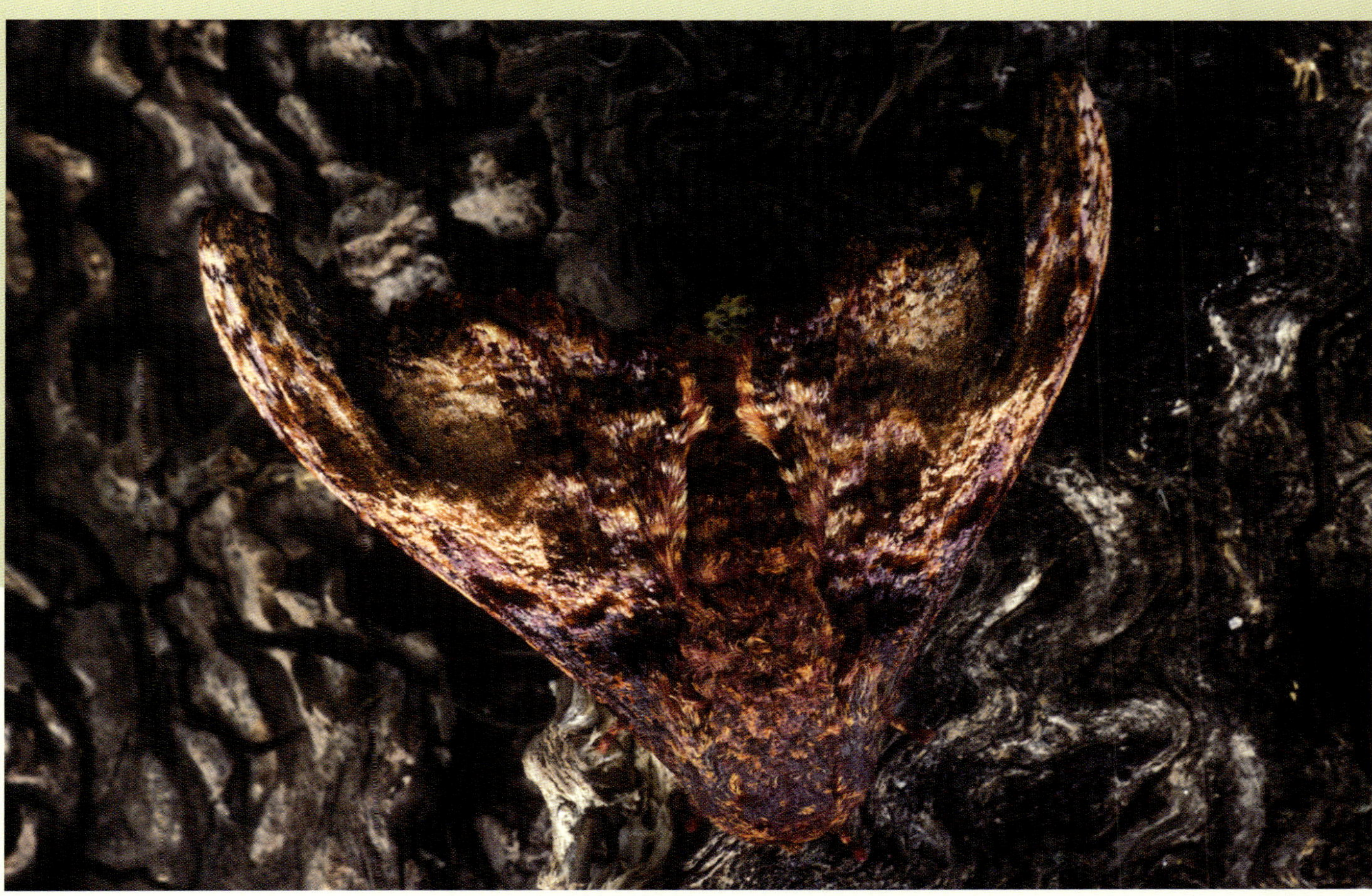

that of a bird with spread wings and tail fan. The hidden bird image is thrown into relief by the more reflective violaceous areas of the wings, more easily seen by adjusting the colour balance. Viewed under near ultraviolet light these areas reflect light more strongly, as birds will see them, bringing the bird wing image into stronger relief. As noted in Chapter 1, every nestling will become imprinted on the image of the parent bird landing on the nest, and so adult birds are likely to react to this image and not in the first instance see a moth.

Another example of bird wings and tail fan is found in *Copaxa denda*, in which eye-spots are added in as "graffiti" objects. In the related *Eacles* species, the paler areas are used to define an apparent reptilian head, like that of a chamaeleon, viewed from the side, and *Eacles penelope* bears a face mask with the broad fringed lips of the North American Horned toad.

Other patterns visible to birds but not readily to ourselves can be revealed simply by altering the colour balance on images so that the more reflective parts of the wings are brightened: this is probably close to what birds see given their greater sensitivity to light at the violet-ultraviolet end of the spectrum.

Other moths that are obvious mimics of dead leaves on the forest floor have evolved simulacra of rodent-like mammals. The dull-coloured fore-wings conceal the hind wings at rest, but have a small whitish spot in the middle of each of the fore-wings, known to taxonomists as a discal spot. These spots become evident when the moths are viewed in the head-down position; this is because the thorax is a good template for the nose, and the long hairs on the body come to resemble fur (see also *Copaxa sapatoza* p.32). But the subterfuge does not end with simple spots. Generally, these spots become highly reflective when light strikes them at a certain angle, so they stand out, even in dim light, against their background, as do reflective road markings.

above: *Copaxa mazaorum* (male)

right: The same photograph as above but colour-enhanced to show the bird image of wings and tail fan

opposite top: *Citheronia maureillei* (female)

opposite bottom: *Citheronia laocoon* (female) showing bird wing pattern

above: ***Eacles penelope*** (male)

right: Head of a chamae eon. Compare with the *Eacles* fore-wing

opposite top: ***Copaxa denda*** (female)

opposite bottom: ***Copaxa moinieri*** (female). This image is not enhanced but the raptor image will now be more apparent after viewing the previous ones

opposite and above: ***Syssphinx colla*** (male). Countershading gives the illusion of depth, and colour enhancement *(above)* reveals a reptilian head with open mouth

right: ***Omiscodes cinnamomea*** (male). Moth, leaf or mouse?

pages 58/59: ***Dirphia centralis*** (male). Fungal growths?

Very commonly, though, they will reflect light back to the eye only when viewed from a certain angle, so you are unlikely to see this in moths in museum cabinets or in photographs of the spread wings. Fungal and lichenous growths pictured on leaf-like wings are found in species such as *Dirphia centralis*, and even on the bodies of caterpillars.

We also see many examples of disruptive coloration in the caterpillars of saturniid moths. We tend to recognize a moth or a caterpillar by its uniform coloration and characteristic shape, but these features can be disrupted by patches of contrasting colour that draw the eye towards them and break the body surface into separate components. The purpose is to prevent or delay recognition by sight.[4] The results are sometimes extraordinarily beautiful.

Moths that mimic leaves of forest trees present themselves, matching the leaf form, as two-dimensional objects. But moths that are mimics of objects that we normally recognize because they are three-dimensional have evolved images that, by highlighting and shading, produce the illusion of solid objects. The zoologist Hugh Cott carried out pioneering work on camouflage in animals, which became of great value in military

Caterpillar of *Citheronia pseudomexicana*

opposite: **Three examples of disruptive patterning: caterpillars of** *(from top)* ***Rothschildia hopfferi*, *Automeris lauta*** **and** ***Automeris umbrosa*. Note the ways in which the feet are obscured, and the toadstool patterns in *A. lauta***

above: Syssphinx molina caterpillar in its normal resting posture: an example of countershading, with a borderline that because of the darker pigment below it, stands out like a leaf midrib. The metallic gold spots further confuse the caterpillar image

right: Citoica anthonilis (male). Moth, dead flower, or bat?

opposite: Holocerina smilax (male). Bark mimic, with a curvature effect created by countershading of pigment

defence, and more recently has influenced the fashion design industry. An artist of outstanding ability in pen and ink, his art taught him that unequal illumination gave a solid object the appearance of three-dimensionality – even when the background colour and texture were the same as the object itself. We are accustomed to seeing objects lit from above. A horizontal cylinder, whatever the colour, is brighter above and appears gradually darker as you focus on the under surface. We expect caterpillars to be the same, but many caterpillars are more darkly coloured on the back and lighter on the under surface. This countershading makes the creature appear two-dimensional, especially when, as is usually the case, the caterpillar positions itself underneath a twig. Cott explains it this way:

In countershading we have a system of coloration the very opposite of that upon which an artist depends. The artist, by the skilful use of light and shade, creates upon a flat surface the illusion of solidity; nature, on the other hand, by means of countershading, creates upon a rounded surface the illusion of flatness.

The importance of countershading was first shown by de Ruiter[5], who, working with the Nobel laureate Niko Tinbergen, used captive Eurasian Jays (*Garrulus glandarius*) in his experiments. In training experiments they found that the birds could not distinguish between a solid cylinder and a countershaded flat model of a cylinder. This is largely to be expected, because the birds rely upon monocular vision which provides few cues about depth, and hence three-dimensionality, but it demonstrated the potential value of countershading in caterpillars. Tinbergen's jays failed to detect stick caterpillars unless one moved or they trod on one, and would then attack both stick caterpillars (which are not countershaded) and twigs indiscriminately until they were discouraged by the taste of the twigs.

Dysdaemonia boreas (male). A moth with an atypical wing shape

opposite: ***Copiopteryx semiramis*** (male). The tails may mimic whip snakes that hang downwards in trees to intercept prey

Unlike caterpillars that are less apparent when their form is perceived as two-dimensional, the converse is obviously true of those that mimic twigs and of adult moths that are mimics of solid features, such as bark, broken branches, bunches of leaves, flowers, and other animals.

above left and right: ***Pseudodirphia regia*** (male), S. America is a species that resembles the flowers of forest trees such as those of the genus *Couratari* (Lecythidaceae). The countershading of the hind parts of the wings *(left)* confers depth to the image.

left: ***Couratari*** flowers

opposite: ***Psilopygida walkeri*** (male), S. America

pages 68/69: ***Dirphiopsis flora*** (male), S. America. Flower or small mammal?

Bunches of dead leaves hang downwards, and moths that mimic these fold their wings accordingly, but many species of roosting bat also hang with the head downwards, and some moths have an expanded tip to the abdomen. A moth like *Citoica* becomes a chimaera of moth, leaves and bat; and perhaps also of flower buds partly opened, and the countershading of the wings and body contributes to these illusions.

Tinbergen and Harvey Croze[6] trained crows to search for mussel shells with red paint on them, and found that they still went for red cockle shells, less often to shells of hens' eggs and not at all to red razor shells. This simple experiment teaches us that colour is more important to crows than shape and that the shape of prey must be radically different from the norm before the birds discriminate against it. Of course we should not generalise this finding to birds as a whole, but camouflage that includes an atypical wing shape is occasionally encountered, and then the shape is radically different, as in *Dysdaemonia* and *Copiopteryx* species.

Now it is tempting to believe that, from this stage, species have evolved larger and more realistic eye-spots, and that there should be an evolutionary progression which mirrors the steps in the evolution of the vertebrate eye. This, however, would be sophistry; each species is exquisitely tuned in design and behaviour to survive in a particular part of the environment – which biologists call an *evolutionary niche*. Any moth that deviates from the *beau idéal*, or fails to fit perfectly into the mould that nature has provided for it, will have a reduced chance of survival: that is what evolution is all about. We can therefore appreciate the fact that the offspring of a moth that frequents leaf litter and is a rodent mimic will be at increased risk if it has larger eye-spots than the average rodent. There are not likely to be intermediates between eye-spots which represent forest-floor predators and those which resemble the large eyes of owls and other nocturnal predators frequenting the upper storeys of the forest, and then, again, between those from tropical grasslands where cats and antelopes are common. We will examine this issue more closely in the succeeding chapters.

Some of the giant silkmoths resemble the large flowers of rainforest trees, which we do not easily see because they are usually in the forest canopy. One of these moths, *Psilopygida walkeri* from South America, closely resembles an orchid flower. Other species present as somewhat ambiguous. The wings of *Pseudodirphia regia* are modelled like flower petals, but the abdomen is banded and can be flexed to imitate a stinging hymenopteran – bee or wasp.

Dirphiopsis flora has floral features that include representations of coloured petals, stamens, yellow anther heads, and a nectar zone at the base of the petals, but it also has a superficial resemblance to the face of a raccoon.

It is known that some insectivorous birds develop a "search image" for cryptically coloured prey. Great tits, for example, learn to detect camouflaged prey from a distance of half a metre or more, and then make a more detailed inspection at close quarters.

Caterpillars include some of the most weird creatures in nature; many have such exuberant colours and ornamentation that it is difficult to find the caterpillar amongst the bristly coating and delicate embroidery of their bodies, while others, perhaps those with which we are more familiar, are drab, worm-like and without aesthetic appeal.

Projections from the body surface can take extraordinary and bizarre forms. The value of this is not simply that there can be a delay or failure in recognition, but that insectivorous birds can learn the features of their prey and develop a search image, so that forms that do not fit the general search image are liable to be overlooked.

The caterpillars of *Lonomia* species are, like the adult moths, cryptically coloured and generally rest in aggregations on bark where they are well concealed by their markings, but the plant-like branched spines on their backs serve not only as camouflage, but also present a shield containing an extremely powerful venom.

A mysterious epidemic occurred recently in a rural area of Rio Grande do Sul, Brazil.[7] Patients suffered haematoma and gangrene-like symptoms, spreading throughout the body, eventually causing brain haemorrhages which, in several cases, led to fatalities. When a search was done in the local environment for the agent responsible the circumstantial evidence pointed to the caterpillar of *Lonomia obliqua.*

On examination, this curious caterpillar camouflaged itself by means of many plant-like hair-growths all over its body. Each tiny clump of hairs had a sharp, cactus-like spine sticking out which easily and painlessly punctured the skin of a human or predator that touched it. They had found the mysterious assassin ... just a caterpillar. Further studies isolated the agents in the toxins contained in a sack at the base of each spine. It was one of the strongest anti-clotting agents ever found in nature. Along with another protein which attached itself to the surface of the body's cells, this caused all the body's cells to leak like sponges as the blood was unable to clot, until the surrounding tissue was full of "bruised" blood; huge haematomas spread over the surface and the interior organs of the body as well as, eventually, the brain causing massive compression and brain death. In fact, more recent statistics reveal as many as 500 people or more have died as a result of coming into contact with Lonomia.

page 70 left: Caterpillar of *Automeris pomifera*

page 70 right: Caterpillar of the African moon moth, *Argema mimosae*

page 71: The highly toxic caterpillar of *Lonomia achelous*

below: *Automeris larra* caterpillar complementing its luxuriant camouflage and disruptive patterning with a bizarre posture as if it were practising a form of insect yoga

opposite: *Automeris belti zaruma* caterpillar

The typical caterpillars of the bull's eye moths (*Automeris* species) are also festooned with numerous branched spines resembling mosses, or horsetail ferns, as in *A. belti zaruma*, but these completely obliterate the body form of the caterpillars. Again, structures of this kind may not always be solely a disguise. It is a painful experience to handle the caterpillar of *Automeris io*, the bull's eye moth of North America. It is likely that saturniid caterpillars with venomous spines are models for others in the same family that are non-venomous. The mimicry may extend further than that: many Costa Rican butterfly and moth caterpillars have spines similar to those of the silkmoths.[8]

ATTRACTION

Lovely in dye and fan,
A-tremble in shimmering grace,
A moth from her winter swoon
Uplifts her face:

Stares from her glamorous eyes;
Wafts her on plumes like mist;
In ecstasy swirls and sways
To her strange tryst.
– Walter de la Mare

The enticement of moths to a flame has providec many poets and writers with a good metaphor for the dangers of yielding to temptation. Unlike Odysseus who was tied to the mast by his men so that he could avoid the sirens, moths appear to have no means of resisting the hazards of a candle flame. The likely explanation for this is built around a strange paradox. A moth's eyes are orientating it to the darkest part of the visual field. Although scientific measurements show that this is not close to the light, the way that animal (and human) eyes work makes it appear so. The individual retinal cells, or the equivalent light sensitive cells in a moth (the ommatidia of the compound eyes) interact with each other to sharpen the perception of light-dark boundaries. This is known as "lateral inhibition"; sensory cells which receive more light inhibit neighbouring cells that receive less, so a gradient of light becomes more of a sharp step in responsiveness. Hence it appears darker in a halo around a light than it does outside that halo (known as a Mach band) and the moths fly towards it in vain attempts to avoid the light.

The North American robin moth, *Hyalophora cecropia* (female) mating with *H. euryalus* (male) (female on top)

A striking characteristic of saturniid moths is the huge feathery antennae of males, and it is these, rather than eyes, that guide the males to the females. Some scientists with an engineering background have theorised that the fine structure of antennae makes them ideal detectors of infrared emissions (to which our eyes are insensitive) and that this is why moths orientate to candles and light bulbs. There is, however, no very convincing evidence that the antennae have this quasi-visual role.

As we inevitably find when we look at the lives of animals in relation to their environment, in what we know by the shorthand term "ecology", we find each animal enmeshed in a nexus of interactions. The colours of moths, the body size, the daily (circadian) activity rhythms, the longevity of adult moths, and the plants and trees upon which the caterpillars feed are just a few of the features that are interlinked. Let us look at the robin moth, for example. The adults emerge in the spring. The males usually become active first and begin to disperse from their emergence site before females begin "calling". This prevents inbreeding in a local population when the females tend to lay all their eggs within a small area without dispersing themselves. Calling involves the extrusion of a gland at the tip of the female's abdomen from which sexual attractant pheromone is broadcast. The female calls at night usually from the vantage point of the cocoon from which air currents will take the pheromone away in the form of a plume of odour. She has to choose her position carefully: if the wind is too strong the odour trail will be broken up by turbulence; and if it is too weak it will not travel far.

Neither the males nor the females feed: they have only mere vestiges of sucking mouthparts, so the energy reserves of the female are devoted to egg production and laying, and those of the male to searching for air currents bearing the female pheromone. The males of many species are active during the day. They are then, of course, more at risk from visually hunting predators and tend to be smaller and more slender-bodied so they can fly relatively fast. The colours of day-flying males are generally much brighter than those of the females (see Chapter 1), which may help to protect them by serving as warning colours or possibly by making them resemble similarly coloured toxic butterflies or moths in flight.

The disparity in size between males and females can be quite marked and is known as sexual dimorphism, the males generally being smaller and more brightly coloured than the females (see Chapter 1). The female robin moth begins her calling soon after emergence and generally stays near the emergence site. Mating takes place over a period of about twenty-four hours, and the female dies shortly afterwards, her energy reserves having run out. Like a small aircraft, the male has sufficient energy reserves only for its journey in search of a female, the difference being that once it has metabolised

Female robin moth, *Hyalophora cecropia*

A typical example of sexual dimorphism: Male *Vegetia ducalis* with the much larger female

opposite: Photograph by P-H. Fabre from J-H. Fabre's *"Moeurs des Insectes"*, Paris 1918, showing great peacock moths "pilgrims diverted from their goal by the light of a lamp"

pages 80/81: Female great peacock moth, *Saturnia pyri*

the fat and protein in its tissues it is unable to refuel (adult saturniid moths lack a functional proboscis and do not feed). The heavier female cannot afford to prejudice the laying of 350-odd eggs in her body by moving around a lot, and tends to lay her eggs in strings on the nearby foliage; while in captivity, to the satisfaction of breeders, she will lay all her eggs in clumps. After laying, the female will stay clinging to the vegetation for a few days and perhaps die after a short flight, her biological function complete.

The robin moth's reproductive cycle, which is a highly efficient system for maximising the number of eggs at a minimum energy cost, is possible because the caterpillars are very catholic in their food preferences ensuring that the females do not need to expend energy searching for food resources, and neither do the males have to search for the females over large areas. The converse applies to moths that have a more limited range of potential food plants, which may be in areas of high biodiversity. Such moths tend to be small-bodied with highly mobile males that are more brightly coloured than the females and fly by day. As female moths generally call during the night, attraction of males by colour and flight pattern, as in most butterflies, cannot operate except, perhaps, at very close quarters, so the scents released by the female are the most important cues used by males in mate-finding. The early writers, including Moses Harris who published in 1766 *The Aurelian or Natural History of English Insects*, a beautifully illustrated tome famous among entomologists, wrote about the process of "sembling" in moths: "... go out with a live Hen [female moth] in a box which is covered with Gauze or Crepe", he wrote, "... the Cocks will come and attempt to get at the Hen. I have known great numbers taken in one Hour's time". But one of the most inspired descriptions is given by the American author Gene Stratton Porter in her book *Moths of the Limberlost*. The Limberlost was a swampy area in Indiana where Gene lived. She was entranced by the moths she came into contact with, and wrote about the robin moth in a way that conveys the wonder that seizes, repeatedly, all who come into contact with the natural world: "there came to me the most delightful experience of my life. One perfect night during the middle of May, all the world white with tree bloom, touched to radiance with brilliant moonlight; intoxicating with countless blending perfumes, I placed a female Cecropia on the screen of my sleeping-room door and retired. Past midnight I was awakened by soft touches on the screen, faint pullings at the wire. I went to the door and found the porch, orchard, and night-sky alive with Cecropias holding high carnival. I had not supposed there were so many in all this world. From every direction they came floating like birds down the moonbeams ... I carefully removed the female from the door to a window close beside, and stepped on the porch. No doubt I was permeated with the odour of the moth. As I advanced to the top step, ... the exquisite big creatures came swarming around me. I could feel them on my hair, my shoulders, and see them settling on my gown and outstretched hands. Far as I could penetrate the

night-sky more were coming. They settled on the bloom-laden branches, on the porch pillars, on me indiscriminately. I stepped inside the door with one on each hand and five clinging to my gown. … Repeatedly I counted close to a hundred, and then they would so confuse me by flight I could not be sure I was not numbering the same one twice. With eight males, some of them fine large moths, one superb, from which to choose, my female mated with an insistent, frowsy little scrub lacking two feet and having torn and ragged wings. I needed no surer proof that she had very dim vision."

It is not certain when Gene Stratton Porter wrote this description, but *Moths of the Limberlost* was first published in 1912. About the same time[1] the writings of the French entomologist, Jean-Henri Fabre were becoming well known and he had a similar experience of moths, which was first published in English in 1919 after his death in 1915. After a career as a school teacher, he devoted his retirement to the study of the insects in his garden in the village of Sérignan, near Avignon in southern France. He wrote about their behaviour in poetic language, carrying out simple experiments that anticipated the findings that biologists such as the animal behaviourist and Nobel prize-winner Niko Tinbergen made over fifty years later.

In his lifetime Fabre had an outstanding reputation as a teacher and scientist: there was a campaign to nominate him for a Nobel Prize[2] and his meticulous observations caused Charles Darwin to describe him as "the incomparable observer". By carrying out simple experiments he effectively discovered the sexual scents now known as pheromones,[3] and this was prompted by what he called a memorable night, "the night of the great peacock". This moth, *Saturnia pyri*, is the largest moth in Europe. The night began with the emergence of a female moth, which he placed under a cloche with a metal mesh. At nine in the evening his young son Paul-Henri ran around the room in great excitement calling: "Come quick. Come and see the moths – big as birds – the room is full of them!" The great peacock seemed to have taken over the house. There were twenty or more flying round the cloche. Over a period of eight nights, a hundred and fifty male moths came into the house attracted to the cage where the female was incarcerated – a number Fabre described as "stupefying" considering the difficulty of finding the moth in the countryside around.

By simple experiments on this and other moths, Fabre deduced that the males found their way to the female by following an aerial scent trail. In all silkmoths the organs of smell are the antennae, which appear rather like bracken fronds in the male; the numerous double-sided branches are coated with microscopic conical sense organs – over 120,000 on each antenna of *Antheraea polyphemus,* for example. The antennae filter the air, trapping molecules of the female attractant, or sexual pheromone, and about 80% of those sense organs respond only to the pheromone of the same species.

The mystery that Fabre strove to understand is explained by the extraordinary sensitivity of the sense organs of the male moths. Very little pheromone is liberated by females. The first moth pheromone to be identified (in 1959) was that of the common silkworm, *Bombyx mori*, and this required an extract of a quarter of a million female moths to determine the chemical structure. The male sense cells in each ultramicroscopic sensory cone can detect individual molecules, although it may take more than a hundred molecular strikes to trigger a nerve impulse. Male moths detecting the pheromone then fly upwind, zig-zagging to keep within the "plume" of scent, which is carried downwind in turbulent air like the smoke from a factory chimney. If the wind conditions are right and turbulence does not break up the odour plume, it will act as a guide to them over a distance of several kilometres at least; it is reported that males of the North American promethea moth, *Callosamia promethea*, have been attracted to calling females from a distance of 33 km.[4]

However, a mystery remained. It had been known for some time that the female scents were not attractive exclusively to the species that produced them. There were reports[5] that moths from different hemispheres could be attracted towards each other and mate. The Curator of Insects at London Zoo found that a male *Samia cynthia* (Asian) was attracted to and mated with a female *Cricula andrei* and a male *Loepa katinka* (China and North India) paired with *Antheraea polyphemus* (North American).

In the late nineteenth century, following the publication of Darwin's *Origin of Species*, various people tried to produce new species by hybridising saturniid silkmoths. One of the most obsessive experimenters was a German biologist, Max Standfuss, who, with typical Teutonic thoroughness, bred an astonishing total of 4,000 hybrids of the three European species, the great peacock moth, the emperor moth, and a close relative, the sloe emperor moth (*Saturnia spini*), the latter mainly confined in distribution to Eastern Europe. All the hybrids were given names and regarded as artificially produced subspecies. One, described as "a large powerful insect" was named in honour of its creator *Saturnia* hybr. *standfussi*. These results, of course, beg the question of why hybrids do not exist in nature, or if they do are extremely rare and valuable collectors items.

Later, when the constitution of the pheromones of many silkmoths had been determined and electrophysiological techniques had been developed for recording the activity of the odour receptors on the male antennae, Ernst Priesner in Germany investigated the overlap of responsiveness in 104 species of silkmoth with the strange result that the antennal

sense organs of closely related species responded to pheromone components of other species in the same genus. For example, the male antennae of three species of *Hyalophora*, including the robin and ceanothus moths, all responded strongly to the same mutual sex pheromone. Six species of *Automeris* (bull's eye moths) all overlapped in responsiveness with each other and with two *Hyperchiria* species, which are closely related. One of the oddest findings was that the great peacock moth of Europe appeared to share the same communication channel with two Rothschildias, the South American species *R. aurota* and the obsidian knife moth (*R. orizaba*) from Central America. The evidence from DNA studies helps to explain this.[6] Biologists are generally agreed that the family of giant silkmoths (Saturniidae) began their evolutionary development in the New World and then spread to the Old World in the eras when land bridges existed. This is supported by DNA analysis. That being so, it is astonishing how much their wing patterns differ, presumably because of adaptation to a different spectrum or predators. We are left with the paradox that in all these cases here there seem to be no barriers to prevent one species from being attracted to a related one and therefore giving rise to possible hybrids. This was counter-intuitive to students of evolution who expected to see pheromones acting as instruments to isolate species. But other researchers were finding that pheromones were rarely single compounds, but almost always mixtures with one or more subsidiary ones, and it was these secondary components that inhibited attraction and courtship of closely related species.

page 83: Reproductive cycle of *Rothschildia aurota*, from South America which, on the basis of DNA analysis and pheromone chemistry, is closely related to the Old World great peacock moth. From *"Maria Sybilla Meriaen. Over de Voorttling en wonderbaerlyke veranderingen der Surinaamsche Insecten (Metamorphosis Insectorum Surinamensium)"*, 1730

above and opposite: Two species that share the same sexual pheromone: *Hemileuca burnsi* (male) and *(opposite) Hemileuca electra electra* (male). (See also Chapter 1)

page 86: Antennae of a male tussah moth, *Antheraea mylitta*

page 87: Antennae of a female *Antheraea mylitta*

Among the Hemileucas that are found in the western United States, Burns' buckmoth has a sexual communication system based on five volatile compounds, and all five must be present in the correct proportions to achieve maximum attractive power for the males.[7] Burn's buckmoth males are attracted to *Hemileuca electra* females, but the species live in different areas and do not normally meet. *Hemileuca electra* exists as two subspecies in the Mojave Desert but these two are separated by different pheromone component ratios. It appears that the sex pheromone of the *mojavensis* moth has been modified in the course of evolution, which might in the future lead to two completely separate species – which do not hybridise at all.

In addition, it was found by Ring Cardé and others that there are many ecological barriers to mating of closely related species. They may be separated geographically – the peacock moth and

In the tropics, where many species of saturniids come to the lights, different species arrive in waves at different times of the night.

On one occasion I took some refrigerated virgin females of the rare Copaxa bella to Peru to try and capture some males. These fly at noon, when the females normally emit pheromone during a period of 60-90 minutes. One female had emerged five days previously and appeared to be getting weak and desperate. She continued to "call" with her pheromone until evening. Shortly after dark three males of the extremely rare Copaxa herbuloti were attracted to the nearby lights and collected. Until that time, only one male of herbuloti had ever been collected: back in the 1800s but not described until 1971 by Lemaire. A local insect collector using our same equipment had collected at the same exact location for five years and had never caught a male. So we are fairly confident that the Copaxa herbuloti males were attracted to the pheromone of Copaxa bella.

Another time I wanted to try to hybridize two Californian moths, Saturnia walterorum, a rare diurnal moth that flies in the spring, and Saturnia albofasciata, which flies in the fall and whose male is also diurnal. S. walterorum flies in the midmorning and albofasciata flies in the late afternoon. So to have simultaneous emergences and mating behaviours I had to refrigerate the cocoons of the walterorum for five months, then for one month I put them into a dark closet with a light-bulb on a time clock set for the light to illuminate a twelve-hour day beginning eight hours later than the actual daytime. Both species normally emerge on a sunny day with higher-than-normal atmospheric pressure, so it is possible to know from dawn whether the moths will emerge that day. When the high pressure day arrived I removed the walterorum from the closet in the afternoon and they soon emerged along with the albofasciata so I was able to confine males of walterorum with females of albofasciata and vice versa. The two species are very different in size but similar in coloration. Males of both species valiantly tried to copulate with their female partners, but the large males of walterorum were unable to penetrate the tiny females of albofasciata and the tiny males of these were ineffectual in mating with the large walterorum, so the hybridization was ultimately unsuccessful. Nevertheless, it was an interesting experiment.

– Kirby Wolfe

the obsidian knife moth are obvious examples – or may be tied to a particular habitat: desert, grassland, forest, etc, by food plant and other factors. Cardé also found that activity periods were a key factor in keeping species apart. Generally, a particular species of moth is active only at a limited period of the day or night, and then only for maybe around an hour. The activity is controlled by a circadian clock, which is set internally by the light/dark cycle and functions as a physiological alarm clock.

The ecological separation into different niches thus involves the period of activity, the season during which the species is present in adult form. It is not just the specificity of the female pheromone that ensures that species remain separate. However we can begin to see how new species may evolve as new populations colonise new areas and become isolated from the rest. Fabre's legacy was thus to awaken the curiosity of biologists and chemists, leading eventually to the commercial use of pheromones on a large scale for management of insect pests. However, very few entomologists today, accustomed to studying insect behaviour in the laboratory or museum, will have experienced the joy and wonder that descended on Gene Stratton Porter, and on the elderly Fabre on another continent when his house was invaded with a cloud of these huge and beautiful moths. At the end of his life in 1915 he had been sufficiently inspired by his experience of nature to say "After eighty-seven years of thought and observation, I say not merely that I believe in God – I can even say that I see Him."

PEACOCKS & EMPERORS

Argus lay dead; so many eyes, so bright quenched, and all hundred shrouded in one night. Saturnia [Hera] retrieved those eyes to set in place among the feathers of her bird [the peacock] and filled his tail with starry jewels. – Ovid, Metamorphoses

Ovid tells the story of Io, the daughter of Inachus, a beautiful princess who was seduced by Zeus. Hera, who was his wife, discovered what had transpired and turned her rival into a heifer, which she placed under the care of Argus. Argus had a hundred eyes around his head, two of which were always open, watching. Zeus, pained at the suffering of Io, sent his son Hermes to kill Argus, whom he outwitted by playing the reed-pipe to close all the eyes in sleep.

When Hera saw the dead Argus, she took his eyes and placed them on the feathers of her bird, the peacock, covering its tail with jewelled stars.[1] Probably the description of "stars" is more apposite than that of "eyes". The eye-spot colours are generated by iridescence, and the light blue surround to the dark centre in particular reflects a high percentage of the light falling on it, including the near ultraviolet to which birds' eyes are particularly sensitive. It has, in fact, been suggested that the peacock's eye-spots (which are less than spherical) are not eye mimics at all, but simply a means of producing high-contrast, eye-catching images in the same way that stripes or banded patterns do.[2]

Male emperor moth (*Eudia pavonia*): two eyes on a wing-scape camouflaged with the colours of heather

Joan Miró's *Morning Star*, 1940

opposite: A convincing image of an owl is easily created in this garden ornament with large circles and a black triangular point

It is perhaps this compact package of highly contrasting colours that draws our attention to eye-like images: eyes and eye-like figures are a magnet to the gaze, as a look at Joan Miró's painting *Morning Star* will demonstrate. The eyes are, in fact, a vital guide to identifying other human beings and other animals.

When we look at a work of art with a human figure in it, our eyes tend to track from the bottom left of the picture up towards the face. This tendency possibly evolved to cope with the threat of encounter with a snake in the path or with a human antagonist. We are endowed with the ability to recognise other individuals, and animals too, by their facial characteristics. Claude Lévi-Strauss suggested that the brain is designed to detect faces the fastest way possible. He argued that primitive man would have needed to distinguish humans from bears and other carnivores, and friend from foe, the consequences of a delay being potentially life threatening.

Political cartoonists portray characters with a few simple strokes of the pen, selecting key features in a similar way to mask-makers. Very often only the size, shape and position of the eyes are sufficient to portray not just a face but one that can be recognised as a particular individual (politician, celebrity or other public figure) with whose image we are familiar.

Neurophysiologists have found that there are "rapid response" neuronal circuits which can detect a face or a face-like figure composed of three circles with a line for the mouth. This is detected by the brain within 130-165 milliseconds (ms). It then takes more than twice as long again to focus accurately on what we see before us and recognise the face. Humans presented with an image of a normal face and one in which the eyes have been displaced to another part of the face take about 700 ms to discriminate between them; monkeys take around 600 ms, and longer still if the face is inverted. Budgerigars identify an image of the face of another budgerigar in 100-200 ms, but if the image is scrambled, with the eyes on another part of the face, recognition takes up to half a second (500 ms). It is clear from all this that once an insect has been detected by a potential predator it has a window of opportunity of little more than half a second to avoid being eaten. A butterfly can become airborne well within this time after having detected the approaching predator. A giant silkmoth, resting on the forest floor or on a tree trunk, similarly has only a very short period in which to respond to a bird investigating it, but its body will not be warm enough for it to fly away; the most rapid response it can make is to slide its wings apart revealing false colours.

The eye-spots of butterflies and moths can have various effects on the behaviour of birds, according to the circumstances. Smaller eye-spots, in particular, can deflect attacks away from the head or add to the verisimilitude of a caterpillar with other snake-like features. Encounters with larger eye-spots can give rise to escape reactions or to mobbing responses that small birds show when they detect a predator in their territory.

In addition to the eyes, the reference point by which we detect a human face is the mouth, for which a straight line suffices, but for a bird or other animal encountering another bird it is likely to be the beak, that is, a vertical line, which when looking at a moth is defined by the body shape of the insect. Replacing a horizontal linear marking with a vertical one placed between two circular eye figures changes the icon of a human face to that of a bird, and if the "eyes" are in the same frontal plane it becomes unequivocally an owl. Experiments with models show that small birds also interpret such images of owls in a similar way.

In the 1950s the British ethologist Robert Hinde carried out some classic experiments on the mobbing response of chaffinches,

which tell us a great deal about how small birds respond to the features of owls. Generally, passerines will avoid live predators but approach and investigate novel or vaguely owl-like objects, even stuffed hawks, weasels, and dog models. Like many other small birds, chaffinches respond to a live predator, such as an owl perched in a tree, by mock attacks, producing special mobbing calls that attract other chaffinches and other species of small bird to the location. They then make repeated short flights towards the predator some of which may result in attacks on it.

Hinde found that even a very crude model of an owl provoked mobbing, but models with two large forward-facing eyes were the most effective. Contributory to the two frontal eyes were an owl-like silhouette, with big head, short neck and tail, solid contours, patterned surface with spots, streaks or bars, brown or grey colour or a combination of these colours, and a beak. Many of these features, of course, are typical of saturniid moths, especially the *Automeris* species.

Robert Altmann[3] described mobbing behaviour of wren tits, blackbirds and Anna's hummingbird, which were directed at a stuffed screech owl, and in another study,[4] 44 separate species of bird were recorded mobbing a screech owl model. The attacks that mobbing birds make on creatures that harass them can be very vicious. Interestingly, Altmann noted that the hummingbird flew around the head of the owl jabbing its beak at the eyes. There are numerous records of corvids attacking humans and animals, evidently

pages 92/93: **_Automeris umbrosa_ (male) with a combination of owl-like features including head shape with well-marked contours, grey colours with streaks, large frontal eyes and a large beak image (outlined in white)**

above: **Peacock 'eye-spot' or 'ocellus' on feathers**

opposite: **Peacock butterfly, _Inachis io_**

page 97: **Owl butterfly, _Caligo memnon_**

in defence of their nesting sites. They commonly attack the head and eyes, and the Australian government has published warnings about these "swooping birds".

Some falcons, hawks and owls, including the American kestrel (*Falco sparverius*), pygmy-owls of the genus *Glaucidium*, and the Eurasian Hobby (*Falco subbuteo*) have two large eye-like markings on the back of their heads which are rarely noticed. Spanish and Canadian researchers[5] claim that these eye-spots can decoy small birds towards the predators, enticing them to mob and fly close to the raptor while it is stationary.

The peacock butterfly was aptly named *Inachis io* (appropriate to the Greek legend) by Linnaeus in 1758. Butterflies generally have small eye-spots, but the peacock is a notable exception. By using experimental models, it has been found that birds are less likely to approach a string of three small eyes than one or two, but the next most effective defence is in the form of two large eyes, and this is the path that very many saturniids have taken.

The peacock butterfly does not fly at night but spends the winter in hibernation in places such as hollow trees where it is perfectly camouflaged with its wings closed revealing only the bark-like undersides. The eye-spots of the peacock butterfly are mainly concentric ring patterns with high colour contrast. At rest the butterfly commonly covers the hind wings with its fore-wings and exposes them as it becomes alert to movement. The hind wing eye-spots in particular show the typical features of vertebrate eyes: a large black pupil with an iridescent bluish corneal reflection, encircled by a white zone such as is seen in the iris of many birds of prey.

The eye-spots of the peacock butterfly are in quite a different setting from that of stuffed or model owls, and the butterfly's wings and hence the eye-spots are in more or less constant motion during the period that it is active. Their function appears to be to startle foraging birds, as David Blest found. Working with Niko Tinbergen, he did simple experiments, first with the small tortoiseshell butterfly, which is brightly coloured but does not have eye-spots, and then with the peacock butterfly. When he rubbed off the brightly coloured scales of the wings of the tortoiseshells he found that they were all attacked and eaten by yellowhammers and reed buntings while about a third of the intact butterflies scared off the birds.[6] The rhythmic opening and closing of the wings of the peacocks and the sudden exposure of the eye-spots were effective in making buntings flee. Intact peacock butterflies repelled buntings 149 times while butterflies with the spots rubbed off deterred them in only 37 encounters. Yellowhammers and great tits were scared on their first encounter with peacocks and thereafter refused to go near even those with the eye-spots removed.

In order to find out more about the deterrent effect of eye-spots, Blest then devised an apparatus that simulated the displays of butterflies: a light box on which he placed a mealworm, with eye-spots or other symbols on either side of it. When a bird approached the mealworm he switched on the light illuminating the symbols. Most escape responses of chaffinches, yellow buntings and great tits occurred in response to eye-spots constituted by concentric circles that were countershaded to appear three-dimensional – in other words there was no way of improving on the effectiveness of the eye-spots that nature had provided for this insect. It was also found that response to eye-like symbols was innate in great tits and chaffinches and did not have to be learnt by experience.

Experiments that have been carried out by other biologists on the peacock butterfly have yielded similar results. But, as we have seen above, the survival of the butterflies, or models, is never one hundred per cent, and the efficiency of the defence varies with the species of bird. Nevertheless, every individual butterfly that escapes predation has an increased chance of breeding and passing on to future generations the genes that helped it to survive.

There is a great deal of evidence, mostly circumstantial, that leads to the conclusion that eye-spots can serve as a form of defence, which may benefit both peacocks and many other species of butterfly when they are at risk from birds foraging for insect prey. Generally, in these circumstances they do not then provoke mobbing or escape reactions in birds but appear to act as false eyes, deflecting pecks to sites away from the heads. Testimony to this lies in butterflies often being found with beak marks on the wings and with a slice of an eye-spot missing. An experiment with mealworms artificially ornamented with small eye-like spots at their extremities has shown that such markings placed on a prey-object, become a focus of attack by yellow buntings.[7]

While investigating the deflection hypothesis, Swedish researchers[8] discovered two unexpected factors that influenced the deflection of pecks: the amount of ultraviolet light reflected and the time of day at which birds forage. They studied the woodland brown butterfly, *Lopinga achine*, which has a row of large eye-spots on its under-wings. These eye-spots have white centres which also reflect ultraviolet light. In the early hours of daylight when many opportunistic birds are most active (the early bird catches the worm), and also at dusk, the colour-sensitive cones in the vertebrate retina cannot operate and the wings are difficult to see because of their dark colours. The reflection from the white centres of the eye-spots decoys the aim of the birds towards them. This is no doubt helped by the sensitivity of birds to ultraviolet light. In brighter light, in which the head of the insect is more clearly discernible, the pecks are directed at the head.

The peacock is not the only butterfly that has eye-spots similar to the eyes of owls; the so-called owl butterflies (*Caligo* species) of the New World tropics have a large eye-spot on the underside of each hind wing and these butterflies are frequently presented in biology textbooks as impressive examples to illustrate mimicry of owls. This, however, is another case of pareidolia. The insect rests with its wings closed over its back and rarely opens its wings to display both eyes together, so it is unlikely to benefit from resemblance to an owl.

The main function of the *Caligo* eye-spots appears to lie in deflection of attacks. Many of these long-lived butterflies are seen with identical bird beak-sized triangular pieces missing, with the apex at the false eye. But the butterfly escapes. However, this may not be the sole function of the eye-spots: it is possible that the *Caligo* butterflies are mimics of poisonous frogs or toads: the eye-spot is coupled with a copy of the tympanum (outer ear) and black band through the eye which are typical of frogs[9] and toads. What we do learn from this example is that it is unwise to jump to the conclusion that large eye-spots signify owls. On the assumption that insectivorous birds are deterred by images resembling larger predators, large eye-spots may have evolved to represent any or a wide selection of nocturnal predators with binocular vision, including members of the cat family, monkeys and other primates.

Huge eye-spots are found on many saturniid moths, and most people believe that they are designed to thwart nocturnal predators. For the reasons given above we need to examine this view more closely, particularly as bats, which do not hunt by sight, can often be observed preying on the moths.[10] Many moths, including tiger moths and hawk-moths, have means of evading bats: they have hearing organs that can detect the ultrasonic cries of an approaching bat. Tiger moths are able to produce ultrasonic clicks that interfere with the echolocation sounds of the bats, and moths of other families equipped with ears perform evasive flight manoeuvres, but saturniid moths have no hearing organs and their main means of avoiding bats is to confine their activity periods to those when there is little or no bat activity.

Ear-less moths avoid bats by appearing at different times in the season. In Ontario it was found that such moths, including saturniids, were most abundant between mid-May and mid-June, before the peak of bat activity in July and August.[11] In the same area it has been shown[12] that most species of the Saturniidae were slow to take flight and more likely to stay put when bats were active: the equivalent of keeping their heads below the parapet.

James Fullard and Nadia Napoleone, in a survey of 84 species of moth[13] came to the conclusion that bats have been a major force in the evolution of butterflies and moths. Bats are highly efficient aerial hunters and Lepidoptera have been able to survive only by flying during the day (i.e. butterflies and a very few moth species) or by restricting their flight activity to periods in the year or periods during the night when there are few bats around. Thus the bull's eye moth (*Automeris io*) flies only in the early evening in North America, at a time when the monstrous regiment of bats is not yet fully mobilized, and is then active for less than an hour.

pages 98/99: ***Automeris grammodes*** (male). Two eyes and a vertical line simulate an owl face, there is also a close similarity to a small mammal such as a shrew

Aglia tau (male). The owl face image is sabotaged (for us) by two extraneous eyes

It is also known that large-bodied moths fly at a relatively higher body temperature of around 40°C, which allows them to fly fast and erratically: another defence against bats. By hugging the ground and flying close to obstacles they manage to keep, almost in a literal sense, below the radar of the bats.

During the roughly twelve hours of daylight in the tropics and up to eighteen hours in temperate regions in the summer, most silkmoths must avoid visually-hunting predators in order to survive, which is achieved, of course, by cryptic coloration and various forms of camouflage activity. Should they be discovered by opportunistic vertebrate predators they are unable to fly away immediately because they need time to reach the minimum body temperature for flight. Their survival then depends upon provoking an escape reaction in any bird or mammal about to eat them. This is achieved by exposing images designed to frighten, particularly copies of the large eyes of a sleeping nocturnal predator that would pose a threat to a smaller insectivorous creature. Such counterfeited images will be more deceiving in the dim light conditions of tropical rainforests.

We can now derive a better explanation for the large eye-spots of the peacock butterfly. This insect hibernates during the north European winter, particularly in hollow trees or rock shelters, which tend to be places where owls nest and sleep during the daytime. In low winter temperatures they cannot fly to escape a predator and hence they, also, rely on the brusque exposure of large eye-spots to frighten birds away. The exposure of eye-spots, which may not be effective against bats, is accompanied by intense ultrasonic clicking sounds, which have been shown to deter long-eared and pipistrelle bats.

The peacock butterfly has a counterpart in the largest European moth (also discussed in Chapter 3); *Saturnia pyri*, known in France by the picturesque name of *Le Grand Paon de Nuit* – literally, "the great peacock of the night" – which resembles it not only in name but in the possession of four eyes, one on each wing. There may be various explanations for the presence of four eyes in silkmoths.

First, in any communication system, repetition reduces the ambiguity of the message. This is all too familiar when we are speaking to one another at a distance, or talking over a phone in an area where reception is poor. The message gets partially drowned in the noise. To telecommunication engineers "noise" means anything that obscures the signal, so in visual communication noise can be smoke, mist, leaves blowing in the wind, shadows or anything similar that obscures the signal. Hence we can assume that four eyes are better than one, literally more eye-catching.

Rhodinia fugax. At close quarters the fore-wing eyespots may be out of focus for small birds

The peacock butterfly and saturniids such as the great peacock moth and *Aglia* species adopt a resting posture in which the fore-wings conceal the hind wings, so they appear as two-eyed creatures, but as their state of alert increases they reveal the hind wing eye-spots as well. This changes the "face" into a four-eyed one. Four small circles within a larger one can no longer be perceived as a face, or at least has the disturbing effect of simulating the sudden onset of double vision.

A more cogent explanation for the presence of four eyes may be that the eyes of most small insectivorous birds are small and have a very short focal length. They have to be quite close to an object – perhaps only a few centimetres away – before they can focus accurately. In addition to that, their field of view may be less than 20°. A bird that sees a moth such as *Rhodinia fugax* with its fore-wings closed over the hind wings exposing only two eye-spots may not be able to see these clearly as it gets closer because they will be in its peripheral, monocular visual field and increasingly out of focus when the moth opens its wings. The closer the bird gets, the more likely it is to perceive the face of a cat.

A smaller species (*Eudia pavonia*) closely related to the great peacock is common in Britain and northern Europe. This was known to Fabre as the *Petit-paon* or "small peacock" and has the English name of the "Emperor moth". The caterpillar of this moth feeds on heather, and when fully grown is green covered with purplish-red tubercles that match the colour of the heather flowers, making it wonderfully camouflaged.
The adult moth, however, practises a different deception. The male is more brightly coloured than the female and flies fast over heathland during the day. The colouring and patterns on the fore-wing of the female in particular closely resemble a grey hare:

left: **Male emperor moth *(Eudia pavonia)*.
Inverting the page reveals an owl face**

below: **Caterpillar of the emperor moth *(Eudia pavonia)***

above: Female emperor moth (*Eudia pavonia*)

right: Brown hare (*Lepus europaeus*)

opposite: The great peacock moth with fore-wing eyespots erased. Bird wings appear and an open beak is perceived more clearly

this animal has a similar white patch around the eye and a crooked white band in front of the eye. In addition, the hair around the base of the wings resembles the hair coat of such mammals. Hares and rabbits, of course, are not uncommon on heathland and the European grey hare has a very similar distribution to that of the emperor moth.

Refer back now to the photograph of the great peacock moth and you will see that the wing patterns, although generally similar, do not suggest rabbits; but ignore or block out the eye-spots[14] and you see pale feather markings and the outline of a bird with outstretched wings. Even more startling is the image seen when the moth is viewed head-down, when it is possible to see the open mouth of an owl or nightjar.

It appears that this form of masquerade or masking mimicry commonly involves the portrayal of animals that are active at the same times of the 24-hour period as the butterfly and moth in question, and are found in the same ecological zone. The caterpillar of the great peacock moth has a predilection for almond trees, of which the moth is sometimes a pest, but these do not normally support populations of hares or rabbits, so mimicry of an insectivore that is active at dusk when the moths fly would seem to be a good option. Rabbits and hares are mainly day-active as is the emperor moth.

UNSEEING EYES

The possibilities of existence run so deeply into the extravagant that there is scarcely any conception too extraordinary for Nature to realise – Louis Agassiz 1807-1873

While rodents and other small vertebrates that forage amongst the leaf litter mainly have small eyes and are myopic, the same is obviously not true of the larger creatures, such as owls, tarsiers, potoos and many nocturnal primates. Owls have large eyes which collect the maximum of light as their circular pupils dilate. The eyes face forwards, which means that they have binocular vision – and it is this that tempts us to think of them as wise. The eyes, though, are fixed in position and to change their direction of focus owls must rotate their heads. This is enabled by a flexible neck that has twice as many vertebrae as we have and means that they can scan an arc around them of some 270°. The main result is that if an owl is facing you then you always see two eyes. Animals that are hunted (principally herbivores) have been able to sacrifice binocular vision in order to have all-round vision without moving their heads – pigeons and antelopes, for example, are alert to any movement in the vicinity while they are feeding, which is possible because they have eyes on the sides of their heads. For a prey animal, seeing a pair of eyes immediately signifies a threat. We also see a pair of staring eyes as a threat, but we are not alone: animals, from dogs to apes use a fixed stare as a threat in social interactions.

Automeris complicata hind wings

In his book *The Expression of Emotions in Man and Animals*, Charles Darwin noted that threat and aggressive intent in cats, dogs and rhesus monkeys were associated with pupil dilation, baring of the teeth, and raising of hair with changes in posture that made the animal suddenly appear larger. This bluff is now termed a "deimatic response" and is a type of defence found in many animals, not least among which are the silkmoths.

The deimatic response of many *Automeris* moths is very similar to that of a carnivore. A moth such as the bull's eye moth (*Automeris naranja*) has a small profile when at rest amongst leaves, with a spot on each fore-wing that hints at the presence of a small eye. When disturbed, it opens its wings so there appears to be a sudden rapid increase in size, made more evident by rocking movements, a sudden display of bright colours, and the miniscule eye-spots are replaced by large eyes with dilated pupils. There are even white markings at the base of the fore-wings that suggest false eyes when the wings are folded close against the body, and teeth or a beak when the moth splays out its wings. But the moth goes further than a cat or dog is capable of doing: it changes its apparent identity at the same time from a dead leaf to (as most people accept) an owl.

Opening of the wings is often only the beginning of the display in *Automeris* moths.[1] This is often followed by falling to the ground, upon which they perform a kind of dance. The wings are pushed right forwards and then flicked down rhythmically, so that the insects hop forwards. Some species perform this dance while making the wings shiver at the same time, and in some the flicks can be so powerful that the moth flips over on its back. The polyphemus moth, if strongly molested will go into as many as two hundred rhythmic repetitions, at a rate of around one per second, of its dance step. Viewing this choreography from above has nowhere near the same impact as viewing it from in front of the moth at the same level, when the phantom owl appears to be making repeated pecking movements.

Owls have a retina packed with rods, which are not colour-sensitive but able to detect very low amounts of light; they are long-sighted. The density of the rods means that the retinal image is ten times as sharp in owls as in our retina. Binocular vision makes it easier for owls to determine the size and distance an object is away from them, an ability they enhance by moving the head from side to side making use of parallax: the apparent shift in position of an image relative to the background that occurs when you move your head and which helps to determine the distance of the object on which you are focusing.

These characteristics of owls also apply to mammalian nocturnal predators. "Seeing in the dark", the ability to resolve and locate the position of small or faint images, brings with it the inevitable constraints of large eyes, forward facing, with a large circular cornea, which

Antheraea polyphemus

opposite: Deimatic display of a cat involving change of form, apparent increase in size, and baring of teeth. 'Cat terrified at a dog' from Darwin's *The Expression of the Emotions in Man and Animals*

lack good colour vision. It goes without saying that these large eyes that can present huge black pupils surrounded by a ring of colour are good features for prey animals to mimic.

However, the coloured eye-spots are visible only to animals in daylight, so we are driven to the conclusion that the red-yellow coloured eye-spots of many moths are most effective against diurnal predators.

We must qualify this by taking into account the so-called "Purkinje Shift". In low light conditions, the sensitivity of the cones of the retina shifts towards the blue end of the spectrum, and in consequence red pigmentation starts to appear black, while we may still be able to detect the blue pigments. In a flower garden at dusk, the colours of the blue and yellow flowers begin to stand out, while the reds become less visible.

During the day-time, resting moths that are well-camouflaged when their fore-wings are folded together thus intimidate predators that discover them by what is known as "flash

coloration", suddenly revealing red orange or yellow contrasting colours against which eye-spots may be set. However, this will not be very effective at dusk or in dim light.

We find that some moths that have flash coloration have smaller eye-spots than those which rely entirely upon eye simulacra. The reason for this may be that predators hunting during the daylight hours do have smaller eyes and pupils than those that forage at night. However, the combination of flash colours with eye-spots, whatever the size, produces a remarkable synergy, as in *Automeris egeus* which, when seen in the head down position, reveals two canine teeth, or the outlines of a beak. *A. larra*, a very similar species, has the black eyes and imitations of the characteristic crown of feathers of the American Harpy eagle (also discernible in *A. egeus*).

The eyelids of owls, as in birds and reptiles in general, close to an almond shape, usually paler in colour than the surrounding plumage or scales. The eye opens in owls by a rapid downward movement of the lower lid, which covers most of the eye, revealing the circular eye. A number of species of saturniid moths mimic this action, producing, in effect, a remarkable conjuring trick. The oval eyelid is embroidered on the cryptic fore-wings, amongst circular spots that could be rain-drops or insect feeding holes, and when these are flipped forwards the false eyes appear in the same location.

Think for a moment about the visually-hunting predators searching for insects and other invertebrates in the low light conditions of rainforests. These predators are

above left: ***Automeris naranja.*** **'Mouse' image**

above right: ***Normal resting position***

opposite: ***Deimatic defensive response***

likely to be birds, snakes, monkeys and other primates. With little ability to detect colour, the main cue an insectivorous animal has available for distinguishing between their predators (owls for example) and moths is light reflected from the corneas of the two forward-facing eyes.

The extremely large eyes of owls occupy about 30% of the volume of the head and are fixed in position, unable to rotate in their sockets. They are adapted for maximum gathering of light in low illumination. Leonardo da Vinci wrote in his notebooks:

The eyes of all animals have their pupils adapted to dilate or diminish of their own accord in proportion to the greater or less light of the sun or other luminary. But in birds the variation is much greater: and particularly in nocturnal birds, such as horned owls … in these the pupil dilates in such a way as to occupy nearly the whole eye, or diminishes to the size of a grain of millet, and always preserves the circular form.

The eyes of most species of owl and eagle have a yellow or orange iris that is perfectly spherical. We rarely see owls with the pupil fully dilated because they are so only in very low light. Wildlife photographers prefer good illumination of their subject, which means that the pupil is then constricted. The pupil can contract very rapidly when accommodating; it is powered by striated muscle that has a faster reaction time than the smooth muscle fibres with which the human eye is equipped. Photographers prefer not to use flash, which would

capture an open pupil, because, amongst other distortions it would produce a reflection from the back of the retina and make the photograph look "unnatural".

The main features of owl eyes can be found in many species of bull's eye moth (*Automeris*): a large black pupil with a corneal reflection, surrounded by a yellow-orange iris, set in a black ring formed where the conjunctiva joins the feathered skin surrounding the eye. Owls appear to be one of the most important nocturnal bird predators: there are reports of large numbers of saturniid moth wings, including those of the luna moth (*Actias luna*) beneath an owl's nest. Otherwise there is scant information on vertebrate predators of moths (apart from bats) and neither is there any data on the effectiveness of moth displays in preventing predation.

top: *Automeris amanda tucumana* (male)

above : The same insect as seen in dim light

With their keen eyesight, owls may be able to detect features of silkmoths in moonlight, and the owl-like eye-spots could be sufficient to confuse them about the nature of their potential prey. Other visually-hunting predators of importance are likely to be opportunists, such as rodents and other small mammals. With small eyes these are generally short-sighted and they will tend to detect moths only at close range, at which juncture they may provoke the moths into enacting a defensive display.

top: **Automeris egeus** (male)

above: **Automeris larra** (male), S. America

Many of the bull's eye moths (*Automeris* species) have eye-spots that conform to the huge circular pupil that owls, or indeed mammalian nocturnal insectivores have, beautiful examples being *Automeris excreta* and *A. belti zaruma*.

In these species we see not only the owl-like pupil with the yellow iris dilated almost completely, but the simulation of a corneal reflection. *Automeris macphaili*, for example, appears to be reflecting back a very intense light from the "eyes". The various patterns of light which can be reflected from the vertebrate eye, such as the eye of an owl, appear to have been replicated in the eye-spots of moths by white reflective pigments. This adds to the realism of the image, at least for us and presumably for other vertebrates. It is an illusion that is used in cartoon characters for that same purpose.

Vertebrate eyes actually have four reflective surfaces, giving rise to the so-called "Purkinje" reflections, which come from the inner and outer surfaces of both the cornea and the lens. *A. gabriellae* eye-spots have a barely discernible corneal reflection marking, a small pin-hole in the centre of the "pupil" but one that is highly reflective under ultraviolet light, and the pupil image is not black but coloured with scales reflecting at the violet end of the spectrum. The brightest reflection from the vertebrate eye is from the outer surface of the cornea and is visible in bright moonlight as well as in daylight. Because it reflects from a curved surface, like a convex mirror, the reflected light is seen mainly as a circular pattern or as a crescent of light (e.g. *A. lauta*) depending on the angle at which you view it.

Automeris jucunda has a roughly crescent-shaped highlight in its eye-spots, and interestingly the imitation pupils are elongated so when viewed from a shallow angle, close behind the insect they actually appear circular: this is an example of anamorphosis, an illusion first developed by Leonardo da Vinci in which an object changes its appearance according to the angle of view. Hans Holbein took anamorphism further in his famous painting *The Ambassadors* which hangs in the National Gallery in London. When the painting is viewed by someone standing alongside the picture what was taken initially to be a streak of light in the foreground transforms into the image of a skull.

It is possible that the asymmetry in the eye-spots of this moth has evolved as a means of confusing small birds or mammals, perhaps in the same way that it confuses us. The defensive displays of *Automeris* moths often include rhythmic movements with the wings fully opened, sometimes giving rise to what Blest[2] called "depression flicks" that cause the moth to hop. The effect of this is obviously to create for the close observer a constantly changing sequence of images of eyes.

Automeris liberia, which also has asymmetrical images in its eye-spots, has "pupil" markings that seem to correspond to internal reflections in the eye. Perhaps this simulates light reflecting off the surface of the pecten: this is, as the name suggests, a comb-like flap of tissue projecting from the back of the bird retina towards the lens. It supports the blood vessels supplying the retina which otherwise would overlay and obscure it.

Eye-spots in certain other species show a resemblance, probably fortuitous, to the reflection of stars and moon from a night sky (*Automeris tridens* is an example). We cannot, of course, be sure that the fine details of the corneal reflection add verisimilitude to the eye-spots that mimic owl eyes. However, we should take account of the fact that the markings of cryptically camouflaged moths (and their caterpillars) very often contain extremely fine detailed representations of leaf structures, fungal hyphae, decay marks and feeding holes in leaves etc., that have evidently evolved

pages 116/117: **Automerula auletes** (female)

above: **Automeris excreta**

right: **Automeris belti zaruma** (male)

opposite: Eyes of white-faced scops owl *(Otis leucotis)*

to deceive predators with high resolution close-up vision – i.e. birds. Nature, remember, is weirder than we think: she does not embellish or decorate; errors are soon weeded out; and the smallest details are part of a design that helps survival.

People still puzzle over the transparent "windows" in the wings of silkmoths; some have suggested that the transparent windows help to camouflage the insects. It is known that the "tooth and claw" windows of the Rothschildias are transparent only when viewed at an angle normal to the surface, which is the angle at which we tend to inspect two-dimensional objects. Viewed from other angles we perceive a reflecting membrane and the window we are looking at becomes a solid surface. This is another example of anamorphosis. In the majority of species the window is the central part of an eye-spot and serves as a means of producing a corneal reflection, as in the African species *Nudaurelia dione*.

The tussah silkmoth, *Antheraea pernyi* has large circular windows which are very transparent. As you change your angle of vision by moving to one side or the other you see an anamorphic image: the windows become eyes which open and close. Note that owls can blink in one eye, rather than both together.

above: ***Automeris macphaili*** **(male)**

opposite top: ***Automeris gabriellae*** **(female), central S. America**

opposite bottom: ***Automeris lauta*** **(male)**

page 122 top: ***Automeris jucunda*** **(male). Place the page close to eye level and the eye-spots become circular – an example of an anamorphic image**

page 122 bottom: ***Automeris liberia*** **(male)**

page 123 top: ***Automeris tridens*** **(male)**

page 123 bottom: ***Imbrasia dione*** **(male) with minute windows that mimic corneal reflections. Note the resemblance to the head of a member of the cat family**

above: Tussah silkmoth (*Antheraea mylitta*) viewed at an angle normal to the surface showing 'open' windows

left: Light reflected from an angle gives a 'closed' window

opposite: Most insectivorous birds are myopic with a narrow range of focus and at close range may see only two eye-spots

When the moth rests on a darkened surface the "open" windows can be seen as the large pupil of an eye-spot, while the "closed" window simulates a closed eyelid. Birds have closed eyes while sleeping, but, in common with reptiles and many mammals have a third eyelid, the nictitating membrane, which is translucent and is used to clean and protect the eye. Owls and other raptors tend to blink when closing in on prey, so it is possible that the closed window is a representation of the nictitating membrane. There is a fine vein running across the window, which actually runs at right angles to the main wing veins, and which amplifies the hint of an eyelid.

However, it is possible that the reflective surface is a semblance of the *tapetum*. This is the reflective crystalline structure behind the retina that gathers all the stray light entering the eye of animals (especially nocturnal vertebrates), reflecting it back to the retinal cells and consequently back to any observer. This is then perceived as a glowing object in daylight or as eyeshine on moonlit nights. Like the morpho butterfly wing a very high percentage of light falling on it is reflected back. Put in simple terms, it acts more like a mirror.

Light reflectec from the *tapetum* of the fox eye at night. Eyes of nocturnal predators will appear this way in moonlight or dim light at night

opposite: At close range, a short-sighted bird with a small visual range will see only part of the wings, as in this photograph, and the eyes will appear to open and close as it moves its head

MOTHS *of the* MOON

Because it is of the forest, the brightest light there is to attract it is the glare of the moon … in whose honour these delicate creatures are named, when on fragile wing they hover above his mirrored reflection – "Moths of the Limberlost", Gene Stratton-Porter

In 1852, the Spanish Zoologist Mariano de la Paz Graells dedicated to the Queen of Spain a large and beautiful moth that he had discovered, in these florid terms, "To the august name of H.M. the Queen, Lady Isabel II, I dedicate this magnificent Saturniid, the only representative in Europe of the section to which belong Diana, Luna, Selene, Isis and other Divas less deserving than our own Queen". The ancient goddesses he mentioned had all had their names co-opted into the Latin names of other species of the Saturniidae.

The moth became known as "La Mariposa Isabelina", or Isabel's moth,[1] considered then as now the most beautiful in Europe, but which is also one of the rarest and most threatened species. Because of its similarity to the Indian moon moth, *Actias luna*, it is also known as the Spanish moon moth. The Latin name in use today, *Graellsia isabelae*, is less poetic than its common names, but harbours the history of its discovery in the mountains near Madrid by Graells, who was a leading figure in Spanish zoology.

Eye-spot of the African mcon moth,
Argema mimosae

Male Isabella's moon moth, *Graellsia isabelae*

pages 132/133: **G. isabelae** (female)

The story goes that in 1939 the Queen's Doctor, Juan Mieg, told Graells of a species of moth introduced illegally from America. Graells went into the pine forests between Madrid and Avila to find it and saw his dog, Curicus, playing with a moth that was fluttering around. He realised this was an unknown species and searched during the following months for caterpillars. This was the only sure way of getting a perfect specimen, the original moth itself presumably having been badly damaged by the dog. In the spring of 1943 a magnificent female moth emerged from its cocoon. He took it back to the palace and presented it to the Queen who wore it on her dress at a palace reception and was said to have been so impressed with its beauty that she ordered a necklace with an emerald pendant engraved with an image of the moth. There is now a monument at the place where the moth was discovered.

The fame of the moth soon spread, and it is said that the British Natural History Museum paid 500 pesetas for a moth (probably about £1,000 in today's currency), and a caterpillar was literally worth its weight in gold. In 1909, Sanz, a forester, set up a business exporting specimens to Germany and other countries. This family business closed only in 1960 when fumigation of the pine forests became common practice.

Graells was one of the most distinguished biologists of his time; his flattery of the unpopular Queen appears to have led to royal patronage. He was made Director of the Spanish Society for Acclimatisation, created by the Queen, and later became Director of the Madrid Museum of Natural History and of the Botanic Garden. Scientists in France and Spain at that time were enthusiastically engaged in the introduction of new flora and fauna – everything from cacti to llamas – and it was a little ironic that when Graells published his descriptions of the new moth entomologists in both countries believed that it had been imported by him from America or India. However, further discoveries were made and ultimately, in recognition of his work, the name of the moth was changed to *Graellsia isabelae* two years before Graells died. Later, populations were found in the French Alps, the Pyrenees and Switzerland, as well as in other mountainous regions of Spain. The French duly claimed it as their own, calling it *l'Isabelle de France*.

Reflecting the long periods of isolation in their forest islands, the Isabella moths from different regions differ slightly in their coloration and markings; they have diverged slightly on their evolutionary pathways and become different subspecies. The French subspecies carries the effusive Latin sobriquet of the Glory of Gaul (*galliaegloria*), and the Pyrenean subspecies is called *paradisea*. Without a place in local economies, it is perhaps not surprising that Isabella's moth was not described until the mid-nineteenth century; but another reason is that it is very rare. Its caterpillars feed on pine trees in mountainous areas over 1000 metres above sea level, and the pupae are resistant to the cold. It is thought that the moth may have been widespread in its distribution in the tertiary era before the last ice age, but managed to survive only where islands of pine forest remained uncovered. After the ice retreated the pine forests in southern Europe were gradually surrounded by deciduous forest at lower levels and the moth then became what is known as a relict species – a relic of former epochs. It is found in the mountains around Madrid, in the Pyrénées Orientales, Andalusia, Catalonia and Navarra. In France it was first found by chance in the French Alps in 1922 and a French film called "Le Papillon" was made which is about the friendship that develops between an elderly collector and a young girl on their joint adventure to find the Isabella moth in the Vercors in south-east France.

This graceful insect has a wonderful design on its wings, which is unique amongst silkmoths: a tracery of branching brown lines in which are scattered four eye-spots like glistening coloured oil droplets. This design has captured the imagination of artists and has been used as a decoration on ceramics, but it is in fact highly functional and protects the insect from predators. The brown lines we see are similar to brown pine needles, and they are set on a green background, so they provide camouflage.

In Isabella's moth, as in many other giant silkmoths, the eye-spots provide the ambiguity, because they signify a rather large bird. Remember that a predator hunting through the foliage of a tree does not have the same perspectives as we do with our feet on the ground, and may come across the moth from any angle. Look at the photograph and you see that the hind wings frame a rounded head, with the tails projecting out like a long beak. Turn the page upside down and you again see the body exposed between the wings suggesting a beak. There are also of course four eye-spots. The predator is therefore presented with a puzzle: is it seeing a bird through the apparent screen of pine needles? If so, why are there four eyes? In the precious seconds or even milliseconds that it takes to resolve the enigma, the moth has had time to escape. There is also an anamorphic image in the female based on the white tooth-like segments at the edges of the hind wings. The semblance of a face and open jaw is revealed when you hold the page at eye-level.

However, the biggest threat to the moth lies not in predation but in quite another direction. The most important forestry pest in Spain is the pine processionary moth, which attacks pine forests everywhere in southern Europe. Although this moth sometimes leads to defoliation of large areas it might still be tolerated were it not for the fact that the caterpillars, which live gregariously in silken nests which they spin, produce irritant hairs that are liberated every time they moult. The hairs lodge in the skin and cause rashes and allergic reactions, sometimes necessitating emergency medical treatment when they get into the eyes or respiratory tract. By law in various Mediterranean countries, including Spain, infestations near to urban areas, popular tourist areas and recreational parks have to be controlled.

The answer to the pine processionary problem has been to treat pine forests with insecticide sprays. The microbiological insecticide *Bacillus thuringiensis* (Bt) is commonly used. Even though this often has to be applied by aircraft, the costs are very low. The spray aircraft can cost only a few dollars an acre to operate and the use of the pesticide is often subsidised. Bt attacks all insects – it is the basis of the ingredient that makes genetically modified crops toxic to herbivorous insects – and especially moth caterpillars.

Male luna moth, *Actias luna*

Moon moths have a very wide distribution[2] and all share the same general body form. Apart from the relict populations of Isabella's moth in southern Europe, species occur in Asia from Japan to China and Indonesia, in Africa, and North and Central America. They presumably came from a common ancestor in prehistory which may have evolved in Asia and crossed into North America by the Bering land bridge. Humans are believed to have crossed the land bridge about twelve thousand years ago, but there is fossil evidence of mammals crossing twenty million years ago. Interestingly, the now extinct sabre-toothed cats have a similar geographical distribution to the moon moths, occurring in Europe, Africa, Asia, and North and Central America.

above: ***A. dubernardi*** **(female). The more reflective wings may make it easier for the males to locate her at dusk or in moonlight**

opposite ***Actias dubernardi*** **(male)**

There is a striking similarity in appearance between the Indian moon moth (*Actias selene*) and the American luna moth (*Actias luna*), and breeders have found that the two species will cross-mate and produce eggs that hatch. Perhaps the *Actias* species crossed more recently, and Peigler's theory that cocoon rattles were introduced into the New World with the first colonizers twelve thousand years ago has relevance here. Modern molecular biology studies give us little more information, although they support prevailing theories that all saturniids originated in the New World, then invaded the Old World, and then reinvaded the New World in more recent evolutionary history.[3] In the absence of fossil moths, we can go no further with saturniid evolution although there is evidence from butterflies that supports movement across the Bering land bridge (see Chapter 7).

The external skeleton of insects is made of chitin, a substance that is quite flexible, but which can be hardened and stiffened by a process of tanning, known as sclerotisation, similar to the tanning process that turns animal skins into tough leather, or gives the colour to bark and twigs. The leading edge of the wings of moon moths is stiffened in this way, and seems to have developed as camouflage, appearing as a twig – sometimes

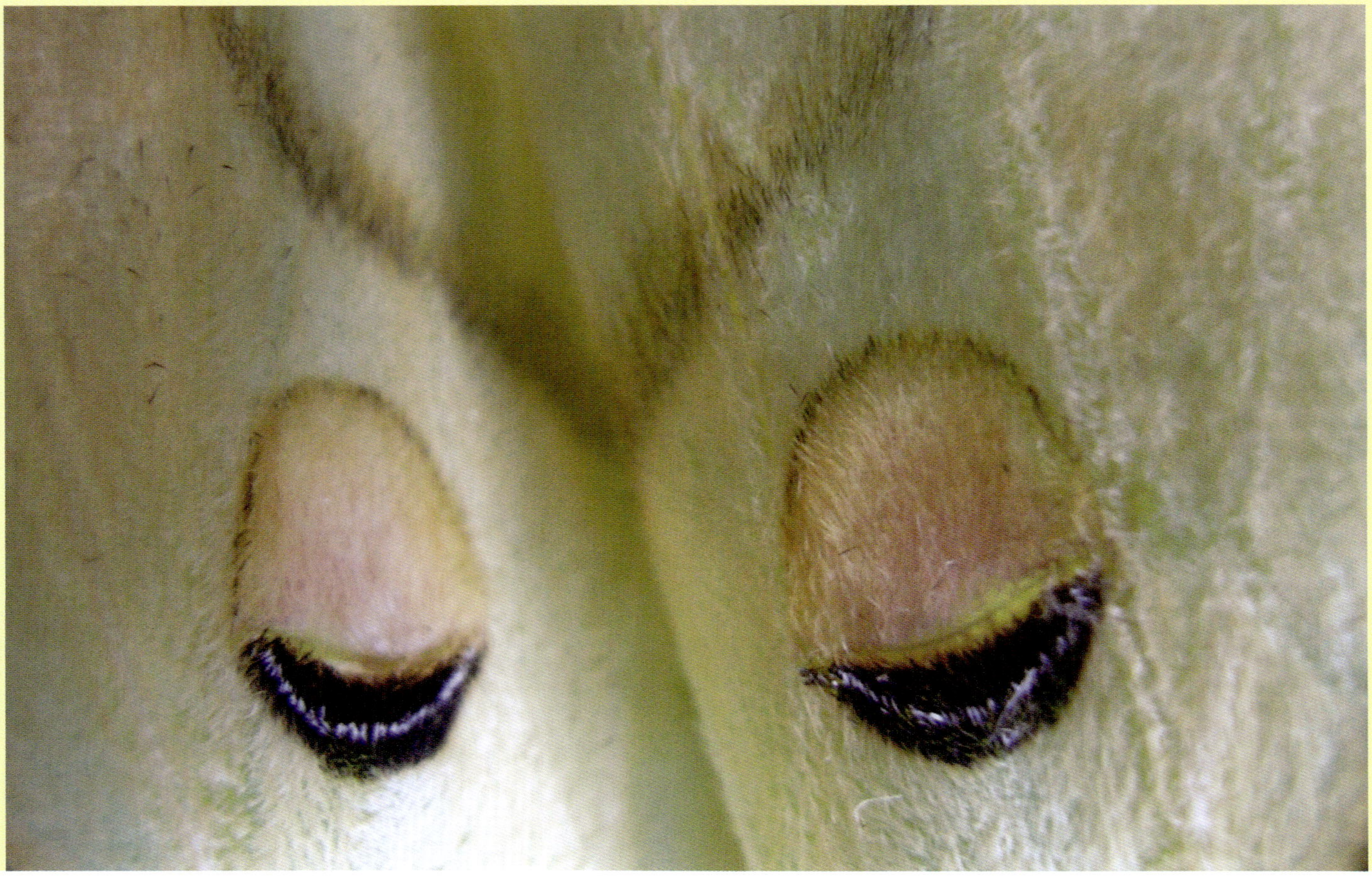

with thorny points on it – against the leaf-coloured wing membranes. As we have already seen, this type of camouflage has been carried to an extreme in Isabella's moth, with the twig or pine-needle motif spread across the whole of the wings.

A variant on the twig theme is found in the American luna moth, *Actias luna*, in which an oval spot on each of the fore-wings resembles a bud or berry attached to the twig by a stem. Some moon moths have curious fore-wing marking like incipient eye-spots. It is tempting to say that these represent "frozen" stages in the evolution of more realistic eye-spots, but similar arguments have frequently been used by biologists in the last century and ignore the fact that animals can remain perfectly adapted to their environments for long periods and are only likely to change their structure and appearance when the environment changes.

The most striking aspect of moon moths is that they are cryptically coloured – either pale green or the brown and yellow colours of dead leaves, but the thick black leading edge to the fore-wings appears, in some species, to be attached to a greater or lesser degree to this. One of the most beautiful insects in the world, *Actias dubernardi*, has only

Indian moon moth, *Actias selene*. Partially closed fore-wings simulate vertebrate eyes and eyelids

Female Madagascan comet moth, *Argema mittrei*

this incipient eye-spot with barely discernible eye-spots on the hind wings in the female. Other *Actias* moon moths have oval eye-spots joined to the fore-wing edge by dark veins, while the Indian moon moth has free-floating oval eye-spots.

The Indian moon moth has front wing eye-spots that no longer resemble a bud or berry, but show an extraordinary resemblance to the eyes of a primate such as an orangutan with half-closed lids. There is even an indication of eye-lashes and a reflection from the lower lid of an eye. This picture is, of course seen only in the bird's-eye-view from in front of the head; when we look at the insect in its head-up position, as we normally do (especially in illustrations), the image is lost to us. The startling effect of similar pseudo-eyes is also found in the African moon moth *Argema mimosae*. On slight lifting of the hind wings the image of an antelope's head suddenly reveals itself.

Careful inspection of the photograph of this moth shows that the fore-wing eyes match those of the hind wings. A startling effect is achieved when the fore-wings, which are folded back at rest, are suddenly moved up to expose the hind-wing

spots which are then in the same orientation, with the "lids" in the upper part of the pseudo-eye. The illusion is again similar in the Madagascan moon moth, where the facial image is perhaps more representative of the circular eyes of the lemurs of Madagascar. Strangely, you may find that the inverted images on these pages lose some of their aesthetic appeal, as if the illusory image intrudes on one that has been accepted as intrinsically beautiful.

It is tempting to compare the tails of moon moths with those of birds. Research recently carried out on barn swallows has shown that the longer tails in the males have evolved partly as a feature important in sexual selection (courtship signals that make the males more attractive to the females) and partly to improve aerodynamic performance. But we have to take into account that swallows have become aerial acrobats and have to be more agile than the insects upon which they feed. If the tail were too long, it might help in courtship but it would also increase the drag and reduce flight speed.

Moon moths, on the other hand, are not predators dependent upon high-speed aerial manoeuvres, and courtship generally takes place in low light and is organised by chemical communication rather than by visual and acoustic signals as it is in birds. When in flight at dusk, the males could be mistaken for late-flying swallows and martins, as the following comment suggests: "[Indian moon moth] males in flight provide a perfectly magnificent spectacle; their movement is of astonishing swiftness and agility, not, in fact, unlike that of a house martin."[4]

There are two important adaptations that make this powerful flight possible. As in a kite, the stiff leading edges to the fore-wings allow them to cut into the wind and create a high-pressure zone underneath the wings and a low-pressure zone above, producing lift. The tails to a kite, which must be very light and flexible, swirl about as the wind catches them and make it stable in winds by increasing the drag. Kites with very long tails can even be flown in hurricane force winds, hence the long tails on saturniid moths would make it easier for them to fly upwind in the odour plume of sexual pheromone released downwind by the female. This may explain why the tails are often shorter in female moon moths, which are more sedentary than the males, spending little time in flight (and, of course, they are heavier that the males because of the eggs they carry).

Daniel Janzen has suggested another function for the tails, which stems from the great susceptibility of moths to predation by bats. Bats hunt using sonar-detecting echoes of the ultrasonic cries they produce, and, as discussed in Chapter 4, are the most important nocturnal predators of moths. As the tails of moon moths swirl around in the wind they will increase the size of the sonar image of what is already a very large insect. This may

The African moon moth, *Argema mimosae*

interfere with the way in which bats recognise potential prey – an echo above a certain size may come from another bat or a bird rather than a moth.

Remember that this is a book that is concerned with ideas and theories, about the way we see things and the way animals do, and none of the suggestions should be taken as fact, but the hypothesis that birds are deceived by ambiguities provides us with a means of understanding the evolution of many patterns including eye-spots. Even a poor imitation of an eye or a face will aid survival of an insect, but as the predator gets close enough to resolve the image the deception will become more fragile. Thus a more realistic eye may tend to evolve. Good examples of eye-spots that might be described as "crude" or "imperfect" are seen in the moon moths (*Actias* species) and many others. These "poor imitations" may have persisted in evolution because they are ambiguous or difficult for a predator to interpret and they thus form excellent defences for these large insects, which are slow to take flight.

What follows from this, of course, is that evolution does not produce organisms that are "imperfect" in the sense that certain forgeries are imperfect copies to our eyes, regarded then as flawed or unfit for purpose. On the contrary: every biologist who delves deep into the study of his or her chosen animal or plant finds that it is "perfect" in the sense that it is exquisitely well-adapted for survival in its particular milieu or ecological niche. But when its environment changes, resources to cope with that may not evolve for very many generations, the species may die out, and be seen unjustly as primitive when it is replaced by a "perfect" one better able to cope with the changed environment. Perfection and beauty, though, are obviously different attributes. The moon moths are some of the most beautiful creatures in nature. Their beauty is something that defies analysis, but their images delight and astonish us. We can only marvel at their extraordinary delicate beauty and grace.

Indian moon moth, *Actias selene* (male)

the ATLAS MOTHS

Nature is not only weirder than we can imagine, it is weirder than we are capable of imagining. – Simon Barnes (paraphrasing J.B.S. Haldane)

In the golden age of Greek mythology, Atlas was one of the race of powerful deities known as the Titans who ruled the world. After the war of the gods, Atlas was banished to the extreme west of the known world (the Atlas mountains) and given the task of holding up the heavens on his shoulders. The Atlas moths are the titans of the insect world: *Attacus atlas* is the largest moth in the world. The largest specimens have a wingspan of around 300 mm (11 in.), but are heavier bodied than *Thysania agrippina*, the Great Owlet moth from the Americas, which has a record wingspan of 310 mm (12 in.).

The common name of the Atlas moth in Cantonese is "snake-head moth". The image of a snake embroidered on the outer edge of the wings is the obvious reason. In Edwards' Atlas moth (*Archaeattacus edwardsii*), which is found in the Himalayas, the snake's head has two eyes (see Chapter 1), rather than a single eye-spot as in *Attacus atlas* (see page 146), so it appears that the creature is turning to face the viewer. The snake image becomes more realistic if the insect is disturbed; it falls to the ground and flaps its wings slowly moving the wing-tips in a figure of eight trajectory, which gives the impression of a writhing snake.

Wing-tip of the ceanothus moth *(Hyalophora euryalus)* with a millipede image showing paired legs on segments marked by wing veins

The "windows" on the wings can readily be interpreted as claws or teeth. They are highly reflective and so appear to stand out in relief against the dull colours of the moth. But look further at the pattern on these wings and you will see a shadowy image that resembles a winged creature – bird or bat – about to alight, or strike with its claws.

The graffiti on the gossamer-like fabric of the wings are representations of the predators that are most dangerous to a small insectivorous bird or mammal: snake or raptor, with dagger-like teeth or claws. You may at this juncture choose not to accept this interpretation and ask, "How do we know that this is not pareidolia: seeing what you want us to see?" There are two answers to this which were examined in Chapter 1. The first applies to mimicry in general. It is counter-intuitive to believe that a human figure could appear in a piece of bread – any explanation would have to invoke the supernatural – but there are obvious benefits for the presence of predator images on moth wings, just as there are for the copying of leaf or bark patterns.

The Giant Atlas moth, *Attacus lorquini* (female), a close relative of *Attacus atlas*. Both have a snake head image with a single eye-spot

Secondly, you will find as we progress that examples of this masking mimicry occur again and again. The comparative approach was the one used by Darwin to establish the concept of evolution by natural selection. He had no experimental proof for his theory but marshalled together numerous examples that established the likelihood that his theory was correct. When we have a phenomenon that we find again and again in different species it becomes obvious that it cannot be one that has arisen as a chance side-effect of gene expression and must be there because it helps survival.

Perhaps even more striking than the Atlas moth is its relative the Hercules moth (*Coscinocera hercules*) which ranges from New Guinea to tropical Australia. This moth is also a contender in the size records – wingspans of over 300 mm (11 in.) have been claimed for some specimens – and has similar features to those of Atlas moths: a snake-like border to the wings, reflective claw images, and a shadowed area suggesting the outstretched wings of a bat. These moths have wing tails, which are short in the female but longer in the male. The long tails are primarily an adaptation

Male Hercules moth, *Coscinocera hercules*

to stabilise flight in turbulent air in the more active male (cf. moon moths, Chapter 6), but here they also produce another snake image. Transverse corrugations simulate the segmentation of a snake and there is an expanded tip to each tail suggesting the diamond-shaped head of a tree snake; and in some specimens there are black and white markings on this "head" that suggest the presence of eyes.

The female has another embedded image. The snake images are less pronounced than in the male and extending towards the edge of the wings are feather tips, marked out in reflective iridescent scales. Fossil remains of flightless feathered dinosaurs, the ancestors of modern birds, have been found recently in northern China. It is tempting to think that this ancient moth evolved in Asia at the time, around the Tertiary period when these dinosaurs with their huge clawed feet and sharp beaks were roaming the continent. The Atlas moths (*Atlas, Archaeoatlas* and *Coscinocera* species) may have survived the great extinction 65 million years ago and become isolated in the Australasian region. If that is so, the question immediately arises, "Why have these features persisted to the present day?" This finds an answer in the fact that natural selection weeds out only features that are disadvantageous in the environment in which the animal lives, including unnecessary characters, the existence of which imposes some cost on the organism.

The novelist Vladimir Nabokov was an ardent collector of butterflies, often weaving references to them in his novels. At one period in his life he carried out research in the Harvard Museum of Comparative Anatomy into the classification of the tribe of blue

Caterpillar of *Archaeoattacus edwardsii*

opposite: ***Rothschildia orizaba*** **(male), Central America**

butterflies known as the Polyommatinae, which are found from Asia to the southernmost parts of the American continent. He then developed a theory that these blues spread from Asia to North America across the Bering land bridge, starting in the Cretaceous period 11 million years ago when insects and flowering plants were beginning to populate the earth, and crossing again in five more waves, until, like the first Americans, they had filtered down to the southern tip of South America. Many years later, Naomi Pierce[1] from the same museum decided to put Nabokov's theory to the test using DNA analysis, publishing her results with co-workers in 2011 showing that it was quite correct. The blues in Patagonia must have crossed Panama into South America 3-5 million years ago.

It seems likely that the saturniids would have also invaded the New World across the Bering Strait, but the general belief among scientists is that they evolved in the American continent and subsequently spread to Asia and the Old World. Now, in the light of Nabokov's theory it seems more likely that the movement to Northern Europe from America would have been a return crossing; and there may have been more than one invasion and return.

The caterpillars of the Atlas moths and their close relatives are equally awe-inspiring. Some instars are covered with fine wax scales and appear frosted as a result. Wax, however, is a useful protection against predatory ants and other insects. The outer skeleton of arthropods has a superficial wax layer that has a natural affinity with other waxes; that is, wax powder sticks to insects. It clogs up their mouthparts, and it covers the tarsal pads on their feet that allow them to grip onto surfaces without slipping. The last stage caterpillar of the Hercules moth is unusual in being blue in colour. It has a conspicuous black spot on each side of its last segment making a false head, while that of Edwards' Atlas moth has a reddish false eye and with the help of backwardly directed spines contrives to look like a small chamaeleon.

In the New World, the Atlas moths are replaced by the Rothschildias. The name of this genus of giant moths comes from Lord Walter Rothschild, an extraordinary character who was obsessed with all things zoological, especially butterflies and moths, and who set up a world-famous museum at Tring in Britain. One of his eccentricities was to harness a zebra to a cart, which he rode around his estate. As an instigator of the Balfour declaration of 1917 he also played a major role in the establishment of Palestine as a home for the Jews.

Walter Rothschild had considerable wealth and employed over four hundred collectors around the world who, in the early part of the twentieth century, helped him amass the largest collection of Lepidoptera in the world: around four million specimens. He took a particular interest in hawk-moths and New World saturniid moths, writing extensive

The goddess *Itzpapalotl* from the *Telleriano-Remensis Codex* and *(right)* from the *Codex Vaticano Latino B* showing features of the plumed serpent god, with the forked tongue of a snake and reptilian jaws

taxonomic works on both groups. Thus one distinct group came to bear his name: the genus *Rothschildia*. He also gave species names to some of these moths.

The Rothschildias are found in a zone from the south-western United States to Argentina. They all share the features of large windows on the wings and wing borders with reptilian-like designs on them. The windows on the wings have the peculiarity that they are transparent when viewed at a 90° angle, vertical to the surface, and hence whatever is behind them is visible. This has led the renowned tropical biologist Daniel Janzen to the conviction that the windows help in camouflaging insects such as *Rothschildia lebeau* while at rest, because they appear as unambiguous images of holes in a rotting leaf.[2] While this is undoubtedly true, the image changes when light falls on the windows. The mirrors reflect back light from certain angles so that you see them suddenly appear to light up as you move past the moth. This is the same phenomenon discussed in Chapter 5 in relation to the circular windows of the tussah silkmoth.

The illusion of flashing images depends upon the fine structure of the surface of the wing membrane. Wing scales are absent on the windows, but stereo electron-microscopy has shown that the gossamer-thick wing membrane is studded with minute projections, like blunt hairs.[3] Think of a brush made of clear glass. Look at it from above and you see the base. Look at it from the side and the light is reflected back from the sides of the glass tufts.

Comparing the windows on the wings with a glass carpet is strangely apt, because the Aztecs were struck by the similarity of the windows of *Rothschildia* moths to their polished knives which were used in excision of the hearts of sacrificial victims and

which were made of obsidian – a glass of fused silica from volcanoes. This gave rise in the Chichimec tribe to the image of the goddess Itzpapalotl, who is represented in the codices – the surviving illustrated manuscripts from the pre-hispanic era – with wings, sacrificial knives and snake images.

Carlos Beutelspacher made a detailed search of images created by the ancient Mexicans[4] and concluded that *Rothschildia orizaba* was the model on which Itzpapalotl was based. It is not difficult to see why. The ground colour of the moth is that of dried blood,[5] sharp pointed knives seem to protrude from its wings, and there is a reptilian image on the wing borders that is vaguely snake-like and very similar to the carved images of Quetzalcoatl, the plumed serpent that adorns the Aztec temples.

pages 152/153: ***Rothschildia hopfferi*** **(male), Brazil**

below: ***Rothschildia erycina*** **(male)**

opposite: ***Rothschildia aurota*** **(female) with bird wing-tip patterns**

The Aztecs of central Mexico were a collection of tribes that settled in the area. Amongst these tribes, butterflies and moths appear to have been important symbolically as images of the soul in departure on death, and in regeneration and new birth. They were also connected with the fire god, and many incensors have

been found in and around the pyramids of Teotihuacan that bear stylized images of swallowtail butterflies which may represent, at one and the same time, soul, flames, and the butterfly goddess Xochiquetzal. Many images also bear circular disc-like structures, occasionally with mica or iron pyrite inserts that resemble the circular windows on the wings of *Antheraea* species that occur in Mexico (see Chapter 5).

The goddess Itzpapalotl was a major deity of the Chichimecs. She is represented in bizarre form in the codices with sacrificial knives, half white and half red, which is the image that the ancient Mexicans must have seen in the *Rothschildia* wings. Like the *Rothschildia* moths she forms a composite image in which other ingredients commonly include (apart from moth wings and antennae) clawed eagle or jaguar feet, eyes, skulls, quetzal feathers, open jaws with a row of teeth, and in some representations, snakes. Most of these are obvious allusions to Quetzalcoatl, the plumed serpent, and it is difficult to avoid the conclusion that original artists saw and recognised the reptilian wing designs.

The legend of Quetzalcoatl derives from a white man who ruled part of the Aztec empire and then sailed away or died at what is today Veracruz. He promised to return or send back his relatives, and it is this legend that greatly influenced the attitude of Montezuma when, in 1518, Hernán Cortés arrived on the Mexican coast. The Aztec ruler welcomed Cortés and was tricked into allowing the Spanish soldiers, whom he believed to be sent by Quetzalcoatl, easy access into the magnificent city of Tenochtitlan (present-day Mexico City).

The remains of the Aztec empire are redolent with images of Quetzalcoatl as the plumed serpent. The similarity of these images, in codices, temple sculptures, etc., with the dragons of China and South-East Asia has been highlighted by Donald Mackenzie[6] who believed the image of the plumed serpent was imported from there, supporting his argument with Buddha-like statues that have been found in early Mexican cultures.

An accumulating body of evidence indicates that elements of Chinese culture may thus have found their way into Pre-Columbian America about a century before the Mexican conquest.[7] To speculate on this: could Quetzalcoatl have been a Chinese shipwreck survivor? Was the plumed serpent derived from dragon images he carried with him or reproduced? What is strange is that the serpent component is not that of a snake – it has crocodile teeth, not fangs. The Chinese dragon, a symbol of power, was based upon the salt-water crocodile. It appears that the symbolic dragon image was conflated with the plumage of the quetzal bird to produce figures like Itzpapalotl, Quetzalcoatl, and other deities.[8]

If we now examine the wing borders of *Rothschildia orizaba*, it is seen to resemble closely the carved images of Quetzalcoatl that adorn the Aztec pyramids, and it also resembles the snout of an alligator or caiman. Certainly it is easier to see a snout with a row of teeth than it is to see a snake, and it may be that the very ambiguity of the image is the main feature of the defence. This image, though, is not the only detail that can be recognised on close inspection of these moths: the whole of the wing pattern in each species can be interpreted as an assemblage of fragments of predator features, each of which may trigger a response in birds at close range.

The obsidian knife butterfly has many obvious similarities to the Hercules moth, and these include a row of feather tips down the centre of each pair of fore- and hind wings, and a dark shadowed inner zone with an outline that simulates a bird or bat hovering or about to pounce. This effect is seen more clearly in *R. aurota* which has wing-tips that provide a perceptual clash with the reptilian figures on the wing borders. It may be that we have an evolutionary compromise here. Depending on the circumstances, transparent or reflective windows, incomplete images of bat or

Rothschildia tucumani **(female)**

pages 158/159: ***Rothschildia zacateca*** **(male). The large hind wing windows may appear as such to a bird a few centimetres away**

bird wings, claws or teeth and reptilian heads may all have survival value in deterring potential predators. Windows in some moths, as Daniel Janzen suggests, may break up the outline of the moths when they are resting on a branch with the wings hanging down, while in other situations they reflect incident light-producing images capable of a different interpretation. However, if they are very large they make it difficult to see the wing silhouette. We can imagine that any small bird, reptile or small mammal that is confronted by this conglomeration of images at close range will be severely taxed to resolve them and decipher the form of an edible moth. Even the image of obsidian blood-stained knives supervened in the perception of the ancient Mexicans, just as the "skull" image on the thorax of the death's head hawk moth becomes a centre of focus and pareidolia for most people that encounter it.

René Magritte: *La condition humaine*. A painting in which a solid object transforms into the scene through a window

opposite: Male ceanothus moth, *Hyalophora euryalus*

In other species of *Rothschildia* the pendulum has swung towards a snake impression and a caterpillar or millipede one. *Rothschildia hopfferi* from Brazil has snake images on its wing borders: the enlarged head on a slender body gives the illusion of a snake with its head directed upwards towards the observer and about to strike.

The wings of *R. erycina* bear a good simulation of a reptilian head and scales, even with sharp-pointed fangs at the tip of the wing, while the wing borders of *R. tucumani* have a more generalised lizard-like appearance with clearly marked rectangular reptilian scales. The snake image is easily recognisable, and from the front, the body and head seem to be those of a large cricket or scorpion, but that image is sabotaged (at least for us) by the extraordinary large windows. This is perhaps another example of satyric mimicry (see Chapter 1) in which a mix of incongruent highly contrasting images is confusing and possibly disturbing to the observer.

This dark inner zone which is present on the wings of many of *Rothschildia* moths is also disrupted by the presence of the windows, which can be very large. The wing patterns of moths such as *R. zacateca* appear quite bizarre because of the enormous windows on the wings. These are larger than those that would normally pass for decay holes in a leaf.

The large hind wing windows can be partly concealed when the insect is at rest making them appear like more giant canine teeth; they give the same impression when the insect lifts its wings partially revealing the underside. Bearing in mind that most insect-eating birds have eyes with short focal length and a narrow angle of vision, it could be suggested that the large hind-wing windows appear transparent to them at close range so that suddenly they are literally looking through a window.

Transparency is a form of camouflage, and in 2000, NASA Techbriefs put forward a proposal for "adaptive camouflage" involving an array of sensors that would project the scene behind an object onto the object itself – an illusion that silkmoths have achieved with their wing design. This kind of illusion, present in some of the paintings of the surrealist artist René Magritte, can be very disturbing to us.

The robin moth and the ceanothus moth of North America are closely related to the Rothschildias and bear very similar scripts on their wings. However, the images on the wing borders are not clearly reptilian and could instead represent millipedes. This may seem a weird suggestion, but there are several reasons why it may be so.

Firstly, the resemblances: millipedes are distinctive in having the ring-like body armour fused together in pairs making what is known as "diplosegments". This means that each apparent segment carries two pairs of short legs. The zoologist Sidnie Manton demonstrated that this arrangement of many short legs gives millipedes the ability to develop great power in their forward movement so that they can push their way through leaf litter and soil – the same principle is used in tracked vehicles. Now look at the figure above and you will see the pairs of short legs that are so characteristic of millipedes, with lines marking the double segments, and the rounded head and large eyes, which in millipedes are actually compound eyes like those of insects.

Male promethea moth, *Callosamia promethea*

opposite: Female promethea moth, *Callosamia promethea*

The North American promethea moth, *Callosamia promethea* also has prominent millipede images on its wing borders. This moth co-exists with the robin moth, but

top: Spicebush swallowtail, *Battus philenor*, the toxic model for the male promethea moth

above: Skull of the sabre-toothed cat, *Smilodon populator*, showing huge canine teeth

right: The American giant millipede, *Narceus americanus*, up to 4 in. (10 cm) in length, which secretes toxic benzoquinones

the males and females of the promethea moth have different defence strategies.[9] The two species are kept apart by the mating period, which is during the night in the robin moth, but the female promethea calls from within the vegetation during the afternoon up until about two hours before sunset: the male flies during this period and not at night, and so is vulnerable to the same predators as butterflies. Its coloration is different from that of the female making it a mimic of the spicebush butterfly, *Battus philenor.* This butterfly is highly poisonous, and its dark wing pattern with white makings down the wing borders is mimicked by several species of non-toxic *Battus* swallowtails – a well-known example of Batesian mimicry (see Chapter 1). Thus the promethea male is part of this mimicry ring; the female, on the other hand, shares her colours with both sexes of the robin moth, and rests like them at night, with her wings folded, vertically among bunches of dead leaves. If these insects are disturbed by touch or vibration of the twigs on which they rest, the male promethea will immediately fly off erratically, like a butterfly, but the female and the robin moths defend themselves with a display of rhythmic flapping movements of the wings.

The millipede image is retained in the male promethean moth as it is in the female. Another possible reason for the avoidance by predators of millipedes is that they do not make a good meal for most. The exoskeletons of millipedes are very tough, impregnated with calcium salts, and approximately circular in cross-section, which gives them great rigidity. This not only helps them to penetrate soil and push their way through it, but it also makes them virtually inedible. When molested they roll into a tight spiral making it difficult for predators to break open to find the head. In addition, many defend themselves with highly toxic secretions including cyanides, which exude from pores on each diplosegment. Other species produce benzoquinones, substances that are highly irritant: they are carcinogenic, kill skin cells, and can cause blindness if they get in the eyes.

Remarkably, some primates, including capuchin monkeys from Venezuela[10] and black lemurs from Madagascar have been observed using millipedes as chemical weapons against mosquitoes and other ectoparasitic flies. Wedge-capped capuchins search for the species *Orthoporus dorsovittatus* in termite mounds and then anoint their fur with the bezoquinone-rich secretions which serve to repel mosquitoes. Sometimes they bite the millipedes first which releases more secretions, to which they appear resistant themselves. This is somewhat analogous to smoking tobacco, because the millipede secretion is toxic and carcinogenic, but it seems that, for the monkeys, the benefits outweigh the risks. It has been recorded that black lemurs that use African giant millipedes become high on the secretions that they use for a similar purpose.

According to *The Guinness Book of World Records*, the African giant black millipede *Archispirostreptus gigas* can grow to 38.5 cm (15.2 in.). In the Upper Carboniferous (340 to 280 million years ago), *Arthropleura* became the largest known land invertebrate of all time, reaching lengths of up to 2.6 m (8 ft. 6 in.). During which geological period *Hyalophora* species evolved is of course a matter for speculation, although we have already noted that saturniids may have been around in the Cretaceous period, but it is perhaps more than coincidental that *Arthropleura* was also a North American species. The other prominent features of *Hyalophora* and *Callosamia* wings are the sharply pointed, dagger-like windows, and these may also represent a North American creature, the sabre-toothed cat (tiger) *Smilodon.* The North American species *Smilodon fatalis* had canine teeth up to 20 cm (approx. 8 in.) long and appears to have been very common prior to the last ice age, ending in 10,000 BC. Large numbers of skeletons have been found in tar deposits in California.

We can now imagine a possible scenario in which the *Rothschildia* ancestors first evolved defences based on copying the appearance of Crocodilia, including alligators and caiman, and of predators with powerful teeth and claws. Some time after the last two ice ages, 60 million years ago or 10,000 years ago, these moths moved northwards with the retreating ice to areas that did not support alligators, and the alligator images became more like those of the toxic giant millipedes. In the meantime some of the South American fauna had evolved some immunity to the secretions of giant millipedes and the pseudo-reptilian mask remained the most effective defence alongside claw images.

the ROAD *from* XANADU

Silk is sumptuous, royal, heavenly, it is exotic, erotic, sensual. Most of all, it is simply sheer beauty. – Philippa Scott[1]

Members of the silk-producing families, the Bombycidae, Saturniidae and Lasciocampidae[2] include the largest moths in the world, the size of small birds, with amazingly beautiful coloured designs on their wings. These resplendent images bring to mind the precious coloured silk dresses, saris and scarves of the Far East, and a world of luxury and transcendent exoticism.

The wings of many species bear huge false eyes on their wings, with large black pupils that have white markings giving the impression of moonlight reflecting back from the surface of a cornea. It is these ringed eyes that have given rise to the Latin family name. The reference may be to the planet Saturn or to the Roman god of agriculture, who is always pictured carrying a large sickle. The planet Saturn is also the embodiment of the Hindu god Shani, the son of the sun god. The sickle symbol of the Roman god is used to represent the planet Saturn, and many of these moths have sickle-shaped translucent "windows" on their wings in place of, or, rarely, in addition to, eye-like images. Presumably, however, the nineteenth-century taxonomist Grote, who introduced the taxonomic grouping *Saturnia,* took the more romantic view, having in mind the planet Saturn with its numinous rings rather than the despised god Saturn, who ate his own children, finally being killed by Jupiter, one who survived. One of the most horrific paintings in existence is Goya's *Saturn devouring his son.*

Male oak silkmoth, *Antheraea harti (pernyi)* – one of the producers of tussah silk, which is still a multi-million dollar business in China and Korea

What unites the silkmoths as a group is the ability of the caterpillars to spin a protective cocoon of silk fibres. The Latin word for silk, *sericum*, derives from *Ser*, the ancient Greek name for China, which is mentioned in the Bible (Revelations). Silk from wild silkmoths has been used in China since at least the early Neolithic period, initially, it is believed, made from species such as the Tussah silkworm, *Antheraea pernyi* and the giant silkworm, *Eriogyna pyretorum*. But silk production in China took a new and crucial turn when the species *Bombyx mori* was created by artificial breeding and was domesticated over seven thousand years ago.

Silk thread is secreted by caterpillars of very many species and used as an escape mechanism: if a caterpillar is dislodged the thread acts like a mountaineer's anchor rope and checks the fall; the insect can then climb back up the thread when the danger has passed. The silk produced by the silkworm when it spins a cocoon is of a somewhat different nature: it is coated with a protein called sericin, which glues the fibres together, hardening the cocoon wall. This glue must be softened to allow the moth to emerge, or to unwind the thread for spinning and reeling. When about to emerge, saturniid moths secrete an enzyme (cocoonase) at the head end of the cocoon where the fibres are spun into rings hardened by sericin. The enzyme dissolves the sericin, thereby softening the fibres and allowing the caterpillar to be reborn as an adult moth from its silken womb.[3] Silkmoth cocoons usually have a thin hard outer and inner wall with loosely packed fibres between them. In many tribal cultures throughout the world they are filled with stones and used as rattles, often placed around the ankles and worn for dances and rituals.[4] The Tarahumara tribe of the southern USA and northern Mexico use *Rothschildia cincta* cocoons

The beauty of silk

opposite: *Bombyx mandarina* (male), the wild ancestor of the domestic silkworm, *Bombyx mori*

pages 170/171: *Copaxa lavendera* (male), Mexico

filled with gravel to make rattles that are worn on the ankles and waist. In this tribe, as in many peoples in the world – notably the ancient Greeks – butterflies and moths symbolise birth, death and the soul.[5]

Richard Peigler believes that the use of cocoons for ritual purposes has been widespread over recent millennia, and that they were probably brought from Asia into the New World by Shamans. Harvesting of cocoons is likely to have been connected with the extraction of silk. Alexander von Humboldt, in his epic travels through South and Central America in the mid-nineteenth century, recorded that wild silk[6] was an article of commerce among the Mixtecs at the time of Montezuma (15th-16th century) and was still traded at that time. Carlos Beutelspacher[7] lists *Copaxa lavendera* as one of the silkmoth species that was important to the Aztecs, although its silk is very brittle.

Silk is a most remarkable substance; the silk from a single cocoon of the common silkmoth, *Bombyx mori*, is a single thread about one kilometre in length (some say up to one mile, others less). It is peculiar in having a triangular cross-section; because of this, the fibres in silk tissue lock together making it possible to produce very thin material for clothing. The plane surfaces reflect light and also have a slight prismatic effect which gives silk its lustrous quality.

There is a growing interest in silk in bioengineering. Combined with biopolymers such as polylactic acid, it makes a stiff fibre, which is used for wound sutures and tissue reconstruction matrices. Both the silk and the polymer, which is synthesised from carbohydrates, are compatible with body tissues and are biodegradable.

Silk is also non-conducting and when rubbed against other non-conductors readily generates an electrostatic charge. In 1759 a certain Mr. Robert Symmer published in the *Philosophical Transactions of the Royal Society* the results of extensive experiments with his silk stockings. He wore two pairs, one black and one white, which produced sparks when he pulled the top layer off. When he pulled them off separately they inflated when held apart, and when they came closer together they were strongly attracted to each other. The entertaining behaviour of his stockings allowed him to deduce some of the basic principles of electrostatics.

Silk threads are very fine but have a high tensile strength, approximately the same as that of stainless steel. The Mongol

warriors of the thirteenth century, when their empire was at its height, wore tunics of silk, usually heavily pleated around the waist, under their armour. This was so densely woven that it was impossible to penetrate with a sword or spear. The coarse silk used, presumably from wild tussah silkmoths, was spun together with metallic threads.

The giant silkworm *Eriogyna pyretorum* is also known as the fish-line silkworm because its silk is tough enough to be used for fishing lines. Although it is found in the wild over a large area of South-East Asia, this moth has been artificially bred since 885 AD. It was introduced to Taiwan for silk production during the Japanese occupation in the first half of the twentieth century. Moth farms were built in central Taiwan to produce silk fishing lines for export. After the moth farms were shut down in 1950 the moths established themselves in the wild.

The common silkmoth, *Bombyx mori*, is not a saturniid but belongs to a very closely related family, the Bombycidae. Evidence from molecular biology suggests that it began as a strain of the wild silkmoth *Bombyx mandarina*. An unknown genius around 7000 years ago, who no doubt was fascinated by insects, saw a practical use for them and experimented in the breeding of *Bombyx mandarina*. The side-effects of the artificial selection for more copious silk production were that the moth has lost its sight, adult pigmentation, and the ability to fly: it has become a domesticated animal as dependent on man for survival as domestic cattle.

The Empress Hsi-Ling-Shih is credited with having founded a new home-based industry for the women of China in 2640 BC, after she dropped a cocoon, fittingly, into her tea and witnessed silk thread disengaging. She became revered as a silkworm goddess, along, it seems with other mythical figures. Long before this, however, silk materials were in use: the earliest is dated to 3000 BC and the earliest fragments of *Bombyx mori* silk date to around 2700 BC.[8] Recent finds in the lower Yangtze river include spinning tools and an ivory cup engraved with a silkmoth, dated to 4-5000 BC.

Subsequent to Hsi-Ling-Shih's discovery, the rulers of China promoted silkworm rearing on a grand scale as an industry in which women could play a major role. Land was given away on condition that part was planted with mulberry, and the export of silkworm eggs or information on the processing of silk was made a capital offence. However, during the pre-Christian period people in other parts of the world were making silk fabric from wild silkmoths, but the further we go into antiquity the more difficult it becomes to determine from which species the silk was made and where it originated.

In about 1500 BC the volcanic island of Thera (now known as Santorini) in the eastern Mediterranean was the site of a cataclysmic eruption. A volume of a hundred cubic kilometres of rock debris was thrown into the stratosphere, four times the quantity ejected

Male common silkworm moth, *Bombyx mori*

in the devastating eruption of Krakatoa in 1883. A 60 m (200 ft.) thick layer of volcanic ash buried the city of Akrotiri, which was inhabited by peoples of the Minoan civilisation centred in Crete, 110 km (70 miles) away. Crete itself was hit by a tsunami wave 35-150 m (115-500 ft.) in height, which, along with contemporaneous earthquakes and volcanic ash, appears to have devastated the northern coastal region of Minoan Crete, including the fabled Palace of King Minos at Knossos, which Sir Arthur Evans began to excavate in 1900.

Excavations of Akrotiri begun by Professor Spyridon Marinatos in 1967 have revealed a treasure-trove of magnificent, detailed wall paintings, similar to those from Knossos and Mycenae, but far more extensive. An astonishing find has been friezes showing fleets of ships with moths or butterflies on the prow and masts, which have been interpreted as symbols of trade in commodities.

The identity of these insects is open to speculation, but it has been pointed out that the antennae are more moth-like than those of butterflies, and that they have eye-spots on

Silkworm cocoons

opposite: The Empress Hsi-Ling-Shih collecting mulberry leaves (from Figuier, *Les Insectes*, 1875)

each wing – a feature not found in butterflies of the region. Suspicion therefore falls on the great peacock moth, which some authors[9] believe was exploited for its silk in pre-Roman times. Another moth that may be represented is the so-called Syrian silkmoth, *Pachypasa otus*, which is found in the Near East and Eastern Europe. Like the common silkworm, *Pachypasa* is not a saturniid but a member of a closely related family, in this case the Lasciocampidae. According to one authority its beautiful white silk competed with that produced by Chinese silkworms until the late nineteenth century.

Controversy surrounds the opinions of Pliny the Elder, who wrote about the Assyrian silkmoth, which was used for women's dresses. "Silkmoths", he wrote, "are also reported to be born in the island of Kos, where vapour out of the ground creates life in the blossom of the cypress, terebinth, ash and oak". "The caterpillars", he adds, "weave webs like spiders, producing a luxurious material called silk, sometimes worn by men". Pliny adds that the processing and weaving of silk were invented by a woman in Kos named Pamphile "who has the undeniable distinction of having devised a plan to produce see-through women's clothing".

Most archaeologists now believe that the Kos silkmoth was *Pachypasa otus*, which has two generations a year as Pliny reported.

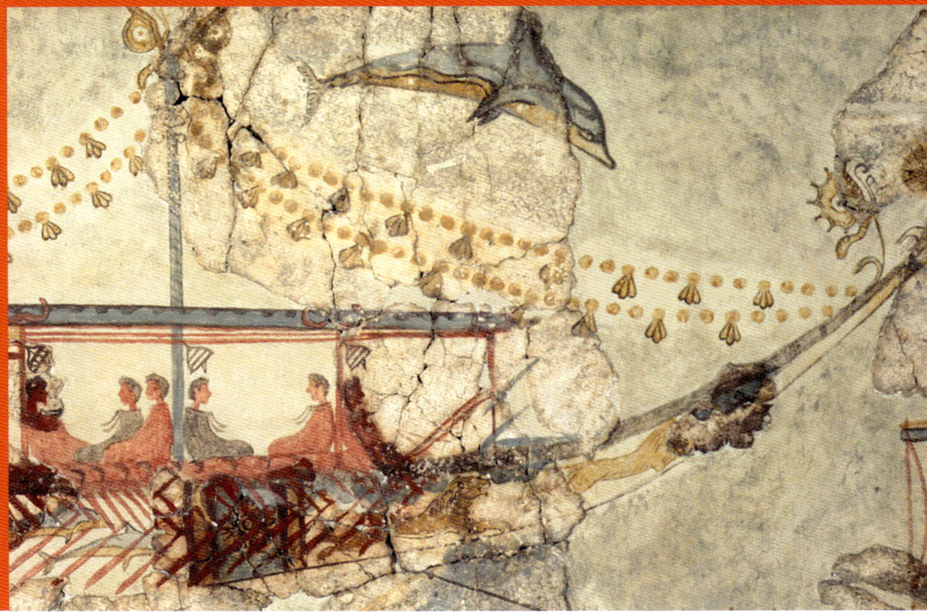

What we do not know with any certainty is which species were present in the eastern Mediterranean three to four thousand years ago. The frescos in Akrotiri show landscapes of palms, with wild cats, antelopes and monkeys. We can only guess at whether insect species from the Indian subcontinent had a range that extended to the eastern Mediterranean. The tussah silkmoth, *Antheraea mylitta*, found in western India, was a sacred insect, the windows in its wings being seen as chakras related to the god Vishnu. The fact that a silk-producing moth with four large "eyes" should acquire sacred status is fascinating enough, but leaves open the questions of whether the moths of the Akrotiri flotilla had a religious significance, and whether they represented the great peacock moth or a moth such as the tussah silkmoth which either had a much larger geographical range than it has today or was imported along trade routes from India. If cocoons were brought into the Mediterranean by traders for processing (there is currently much controversy about which cocoons were used for silk spinning on Kos in ancient times) then it seems likely that moths would have occasionally emerged from some of them.

The history of silk production and its impact on world economies is a fascinating one, which we can touch on only briefly here. The development of technology for mass rearing, extracting and spinning the silk in the Neolithic period was of immense importance for the Chinese nation resulting in the creation of the Silk Road during the Han dynasty (200-220 BC) allowing the Chinese to dominate the world markets for at least two thousand years. The common silkworm and the technology for silk processing were zealously protected: the death penalty applied to anyone taking the insects out of the country.

Moths with eye-spots on the mast and the prow of a ship in the "flotilla" fresco. From Akrotiri, Thera (Santorini), 3000-2000 BC

opposite top: Male great peacock moth, *Saturnia pyri*

opposite bottom: Indian tussah silkmoth, *Antheraea mylitta*

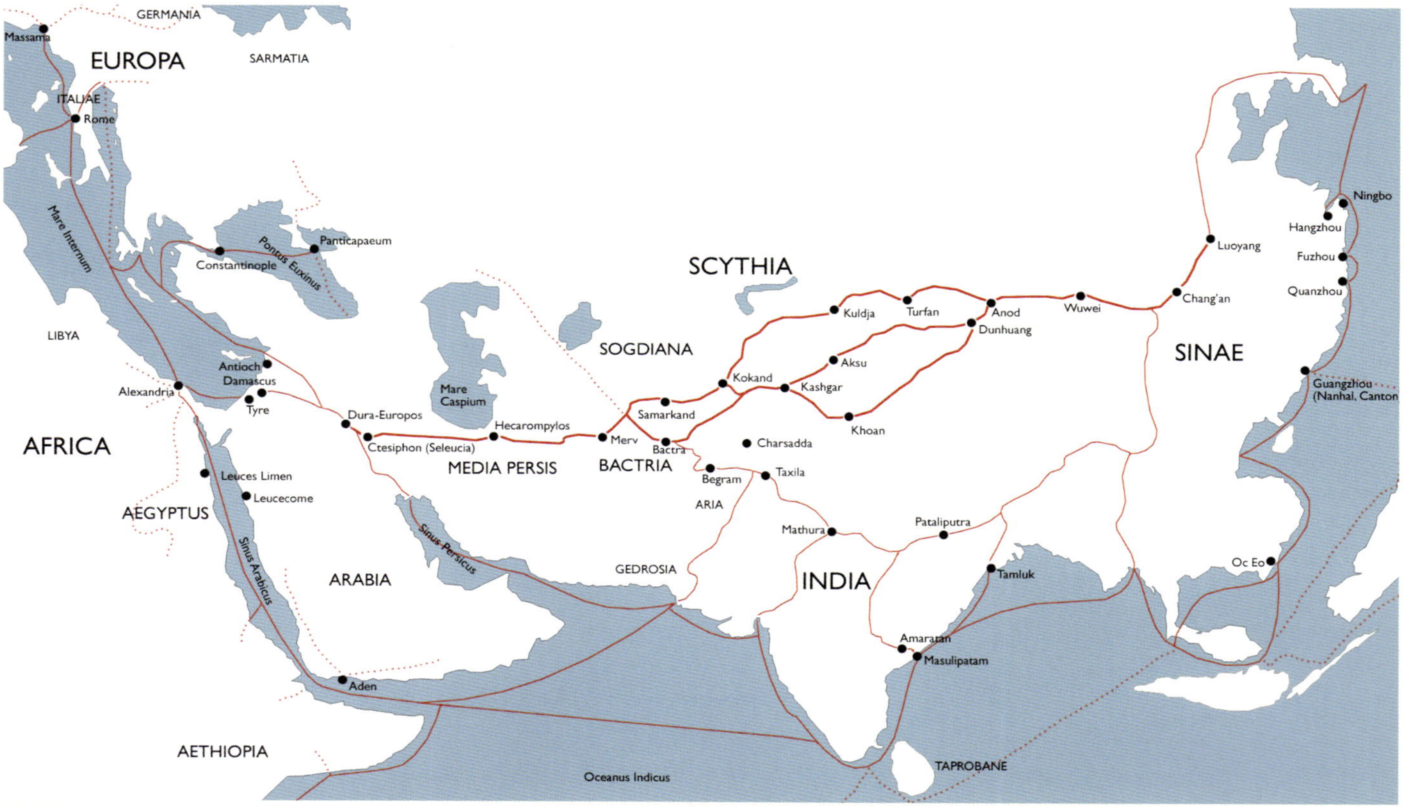

It is even argued that the written Chinese language owes its origins to the silkmoth. As the technology developed in the bronze age it was necessary to record instructions for mass rearing of the silkworm and for the silk extraction processes. Sericulture found its way so thoroughly into the Chinese language that 230 of the 5,000 most common characters of the mandarin "alphabet" have silk as their "key".

Ultimately, *Bombyx mori* was smuggled out of China in AD 552 by two Persian monks. According to one account, the monks "amid their pious occupations ... viewed with a curious eye the common dress of the Chinese, the manufacture of silk, and the myriads of silkworms, whose education, either on trees or in houses, had once been considered the labour of queens". On hearing this, the Emperor Justinian offered a reward for the insects and the way in which they were reared. The monks returned to China and managed to smuggle out silkworm eggs concealed in the hollows of a cane and bring them to Constantinople. From then on "a cane-full of eggs" became the equivalent of a packet of plant seed (for a long time the eggs were called "seeds" or "grain") and parts of the European landscape were transformed by plantations of mulberry trees to support the new industry, which began to play a role in the wealth and destiny of nations. The mystery remains of how eggs were prevented from hatching for the two years of the journey, "being in such interesting company" as one writer put it – if indeed they did so. Nevertheless, from this seed, silkworm culture spread like an epidemic in Europe in the succeeding centuries bringing not disaster but employment and prosperity to millions of people.

The Silk Routes

We are now very familiar with the concept of the "butterfly effect", in which the origin of a hurricane can notionally be traced back to the air set in motion by a butterfly's wing, perhaps on the other side of the world. As China rises again to become one of the dominant world powers we can attribute that, by analogy, to the "moth effect".

Marco Polo set out along the Silk Road in 1271 and found silks already well established in the markets of Persia, China, present-day Iraq and Russia. He arrived at the palace of the Kublai Khan in 1275 and found the walls covered with gold and silver with a hall so large it could accommodate six thousand people. This palace, "a stately pleasure-dome" as the poet Coleridge described it in his famous poem *Kubla Khan*, was made of bamboo cane supported by two hundred silk cords, and was demountable so it could be set up wherever the Emperor wished to be. On reaching Kublai Khan's city of Cambalu (now the site of Beijing), Marco Polo found that "No fewer than 1000 carriages and pack-horses, loaded with raw silk, make their daily entrance into this city: and silks of various textures are manufactured to an immense extent".

The Silk Road is a network of trails and sea routes between China, the Middle East and the Mediterranean, extending over four thousand miles. Once established there, it was extended to the seaborne empires of Spain in the New World and across the Pacific eastwards by the Chinese. It catered not only for the silk trade but also for trade in luxuries such as precious stones and spices, connecting China, India, Tibet, the Philippines, the Middle East, Africa and Rome, thereby setting the scene for the cultural development and politics of the modern world.

Describing the great Chinese fleets that circumnavigated the globe in 1421, a hundred years before Magellan, Gavin Menzies[10] describes the "great sails of red silk, light but immensely strong … like great clouds in the sky". Silks and porcelain were traded in the Pacific, India and Africa.

The dark side of the "moth effect" is epitomised by an outbreak of bubonic plague which was first recorded in China in 224 BC. This spread along the Silk Road in the 1330s and created the first pandemic, resulting in twenty-five million deaths over five years and killing nearly half the population of Europe.

Silkmoth rearing eventually spread to Europe where silk commanded huge prices. At the time of Alexander the Great, lengths of silk became a currency and silk was literally worth its weight in gold. Julius Caesar introduced silk to Rome (although silkworm rearing came only in the sixth century AD). It is recorded that, at a gladiatorial combat in the Colosseum one day, he replaced the tent of coarse cloth normally used to provide shade with a magnificent awning of oriental silk. Thereafter he wore purple silk cloaks when he appeared in public.
The Romans were overawed by Caesar's silk and the patricians then took to wearing silk as a symbol of their wealth and status, until during the reign of Tiberius the Senate forbade the wearing of it because of the economic cost to the State. The wearing of silk clothes by women was considered scandalous: the philosopher Seneca wrote, "The adulteress may be visible through her thin dress, so that her husband has no more acquaintance than any outsider or foreigner with his wife's body".

The desire to colour silk gave rise to another huge industrial enterprise. Silk is a protein, and can be readily stained by indigo dyes, which are then colour fast. The dye of choice for the peoples of the eastern Mediterranean was a purple dye derived from crushed *Murex* and *Stramonita haemastoma* shells, which were common in that region. The dye was in common use by the Phoenicians and known as *Tyrian purple*. Although the principal species, the spiny *Murex brandaris*, is edible, it secretes a compound (an indigo derivative) that is initially colourless but turns red-purple in light. This dye helps to protect the mollusc from predators and the secretion is also placed over its eggs, which are protected by its antibiotic activity. The quantities of dye in a single mollusc are minute, and it is estimated that twelve thousand shells would have supplied sufficient dye (1.4 grams) to colour only the rim of a Roman toga. Because of the expense of production, silk dyed in purple became the ultimate status symbol, reserved for royalty and the aristocracy in Byzantium and Rome, and subsequently incorporated into the vestments of the Christian Church and the British legal profession where it continues to symbolise power, influence and nobility.

It is generally believed that Cleopatra wore silk, and that silk was also used for sails. Shakespeare described Cleopatra's barge as being "like a burnished throne, burned upon the water; the poop was beaten gold, purple the sails . . .". Tyrian purple was as costly as silver: Clytemnestra in Aeschylus's *Agamemnon* says, "There is the sea, ... producing stain of abundant purple, costly as silver and ever fresh, with which to dye our clothes". We have, then, silk worth its weight in gold combined with dye worth its weight in silver – a luxurious harvest seeded by moths.

Venice became prominent in silk manufacture following the trade established by the Romans, and velvet, damask and satin were developed. It is believed that silk first found its way into Britain in 790 AD, when the emperor Charlemagne sent two silk vests to Offa, the king of the ancient province of Mercia. Although France did not develop a successful silk trade until the sixteenth century, the landscapes of Europe, Britain and Ireland were changed in the centuries following by the extensive plantation of mulberry trees, and, particularly in France, Italy and Spain, silk production became a vital element in the economy.

In 1520 a meeting took place between the King of France and the English king, Henry VIII, at what became known as the "Field of the Cloth of Gold". This involved profligate expenditure as each party strove to impress the other with the extravagance of their finery and hospitality.

A pavilion for Henry was set up with 2,800 surrounding tents, using silk drapery woven with gold thread (the cloth of gold). Sixteen thousand sheets of silk were used to make the tents. The noblemen present were richly attired in cloth of gold, velvet and silk costumes. The cost of this magnificence was enormous and placed a severe strain on the finances of both countries. The meeting brought few consequences: no doubt, it was the silk industry that benefited most.

The growth of the silk industry in Europe gathered pace in the eighteenth and nineteenth centuries. As an example, in 1786 the city of Lyons had 15,000 silk looms employing 12,700 people; by 1825 there were 20,101 looms and 8,526 silk factories. At this time too, over 70,000 mulberry trees were planted near Slough (a short distance from London) and 400,000 in the county of Cork in Ireland.

But disaster was waiting in the wings. In the mid-nineteenth century, as silkworm breeding became more intensive, disease broke out among the cultures and silk production fell by around 90% in southern Europe bringing ruin to rural areas. The solution proposed was to import other species from India, which although they produced coarser silk, would not be subject to disease.

In 1858 the ailanthus silkmoth, *Samia cynthia canningi*, was brought to the Jardin des Tuileries in Paris, where it fed on the Tree of Heaven (*Ailanthus altissima*). In 1860 the moth was established in the Jardin d'acclimatation in the Bois de Boulogne where large numbers of caterpillars and cocoons could be seen that summer feeding in the open air on ailanthus trees. Its descendants established themselves in the parks and boulevards, and are still to be found in the environs of Paris.

This moth was also taken to Colchester in England. Lady Dorothy Nevill obtained both the eggs and the ailanthus tree from Australia, where the moth had previously been introduced successfully for silk production (the moth is still to be found in the Sydney area).

In 1864, Alexander Wallace, a friend of Lady Nevill, planted 3,000 ailanthus trees on railway embankments. He calculated that 1,000 cocoons would provide 500 grams (1 lb) of silk, worth £1 when spun. He explained, "all the main labour of worms might be done by women and children of ten years old, so expenses would be trifling".

Samia cynthia was also introduced into the United States about the same time – apparently for other reasons – where the moths established themselves along parts of the eastern seaboard. The silk of the latter was used, among other things, for producing fishing lines.

Thanks to Louis Pasteur, the crisis brought about by the disease of the common silkworm was brought to an end, and after only a few years it became unnecessary to resort to the use of wild silkmoths. In 1865, his friend Dumas persuaded him to look at the disease problem in the village of Alais in southern France. Pasteur was first confused by the fact that there were two pathogens at work, rather than one, as was previously thought: a bacterium causing what is termed "pebrine" disease, and a virus. In 1870 he published a two-volume work in which he proposed that the problem could be solved by removing eggs which showed signs of infection from the pebrine bacterium and unhealthy silkworms suffering from the viral disease. By taking over a silkworm-rearing establishment when the industry was in a ruinous state he managed to achieve a profit of 26 million francs in the first

The Field of the Cloth of Gold engraved by James Basire in 1774 from an oil painting in the collection of George III of England

pages 182/183: Female Ailanthus silkmoth *Samia cynthia*, a producer of eri silk

year (about £3 million). The wild silkmoths had outlived their usefulness when this became widely known.

One variety of the eri silkmoth, regarded by some as a subspecies, *Samia cynthia ricini,* produces a type of tussah silk also known as eri silk, and feeds on the castor oil plant, *Ricinus communis*. The tissues of this plant and especially the seeds contain the toxin ricin, and the plant is considered to be the most poisonous known: six of the seeds are sufficient to kill a horse. *Ricinus* is the Latin word for a tick. The name arose because the seeds, which are egg-shaped with a dark tip section, resemble ticks. Such are the weird creations of nature that the giant prickly stick insect (*Extatosoma*) of Australia, which is amazingly camouflaged to look like a dead leaf, has eggs which mimic the poisonous *Ricinus* seeds. In one of the most notorious assassinations of the cold war in 2003, the Bulgarian dissident Georgi Markov, who worked for the BBC World Service, was walking across Waterloo Bridge in London when he was hit in the foot by a dart loaded with ricin fired from the tip of an umbrella. He is buried in the village of Whitchurch Canonicorum in Dorset, England.

Despite the toxins in the plant, which keep the great majority of insects at bay, caterpillars of the eri silkmoth are able to tolerate it. The principle that "one man's meat is another man's poison" applies to all animals and especially to insects. Once the eri silkmoth had evolved means of detoxifying the ricin it had virtually unique title to the resources the plant provides. However, the moth has been bred for silk production for over 800 years in China, and is now unable to survive in the wild.

One of the most serious forestry pests in North America owes its existence there to silkmoths. The gypsy moth, *Porthetria dispar*, was introduced from its home in Europe to Maryland in the United States in 1869 by the French artist and entrepreneur Étienne Léopold Trouvelot. Trouvelot had already experimented with silk production from the American polyphemus moth (*Antheraea polyphemus*). However this is not a very good silk-producer and he decided that an alternative food plant to mulberry was needed to develop a silk industry based on *Bombyx mori*. Trouvelot planned to hybridise the gypsy moth, whose caterpillar feeds on oaks and other deciduous trees, to produce a silkworm that would feed on the trees in forests of eastern North America. This was a disastrous enterprise. Hybridisation failed but gypsy moth caterpillars escaped and quickly established the species in the wild. After twenty years, the environment began to suffer drastically. Initially, arsenical insecticides and later DDT were used on a large scale, culminating in 1957 in the spraying of three million acres of forest, at which point protests brought the use of DDT to an end. Millions of dollars are still spent annually to stop the perennial spread of this insect across the eastern forests of North America that started nearly a century and a half earlier.

Silkworm production as a cottage industry has existed for thousands of years in the tribal regions of Assam, and has depended on the harvesting of cocoons from the trees on which the eri, muga and tussah silkmoths breed. Often, individual families take ownership of certain trees. The hill tribes of Bengal, according to a publication in 1802, detect the insects by searching for the droppings of the caterpillars on the forest floor. Having located them they keep watch to protect them from predation by birds and monkeys. The richly coloured textiles, which are the fruit of the labours of the women are still produced by hill tribes today using methods that employ locally available materials, including enzymes extracted from plants, and dyes such as cochineal, indigo, madder and turmeric.

Silk from wild silkmoths is generally regarded as inferior to Bombyx silk. It is coarser, and more difficult to spin because the moths break the fibres on emerging from the cocoon with the result that only short lengths can be unwound. Before the cocoons of the common silkworm are unwound the pupae inside are first killed by immersing the cocoons in boiling water or by insertion of a needle. This, however, does not find favour among members of the Jainist and Hindu religions because of their reverence for living creatures. Mahatma Ghandi, for example, would not wear Bombyx silk, and usually wore dhotis of coarse cloth.

China now produces 70% of the world's silk, but because of rapid urbanisation, especially around Shanghai, and the effects of

droughts, there is now less land for mulberry trees. The price of silk is thus increasing rapidly as production slows.

Fine silk today remains a luxury item, associated with high fashion, court and ritual dress. The success of the most famous name in the silk fashion industry, Hermès, was based on their scarves. In 2010 the market value of this company was 18.6 billion euros – higher than the GDP of between fifty and a hundred of the smaller countries of the world, and all achieved with the help of a moth. The eighteenth century poet William Cowper neatly summed up the value of the silkworm to mankind:

Well were it for the world, if all
Who creep about this earthly ball,
Though shorter-lived than most he be,
Were useful in their kind as he.

Antheraea polyphemus (male).
A native American silkmoth

FOOTNOTES

Chapter 1

[1] The Superfamily Bombycoidea is considered to contain six families of moths: Lasiocampidae, Eupterotidae, Brahmaeidae, Sphingidae plus the Saturniidae and Bombycidae. The Lasiocampidae have been important sources of silk for human use for at least many centuries.
[2] In most species the two sexes are similar in colour and fly together either by day or night.
[3] Named after the colours of early police cars in the USA.
[4] Darwin was confused about the role of colours in animals, attributing them to sexual selection until Bates and Wallace gave him their views. He wrote from Malvern Spa where he was undergoing treatment for chronic illness – hence, no doubt, the reference to warming of the blood.
[5] P.E. Howse & J.A. Allen, "Satyric mimicry: The evolution of apparent imperfection", *Proc.Roy.Soc. B*. 257, 111, 1994.
[6] Kirby reports that they are good to eat.
[7] R.L. Smith and W.M. Langley, "Cicada Stress Sound: An Assay of its Effectiveness as a Predator Defense Mechanism", *The South-western Naturalist* Vol. 23, 1978, pp. 187-195.
[8] P.E. Howse *unpublished*.
[9] *Ibid*.
[10] P.E. Howse. *Butterflies: Messages from Psyche*. Also published as *Butterflies: Decoding their Signs and Symbols*, London and New York, 2010.
[11] K. Stapput et al., *Current Biology*, Vol. 20, Issue 14. 8 July 2010, pp. 1259-1262.
[12] P.E. Howse, *ibid*.
[13] Donkeys are similarly marked, although they are descendants of the African wild ass. North American equines became extinct about 9000 years ago and so their colouring in not known.
[14] H.B. Cott, *Looking at Animals: A Zoologist in Africa*, Collins, London, 1975.

Chapter 2

[1] P.E. Howse, *Butterflies: Messages from Psyche*, Papadakis, London, 2010.
[2] Black and white stripes or networks are found in many species of tiger moth.
[3] It was only in the final editing that I suddenly saw this, illustrating that one so often sees what one expects to see. I had hundreds of similar experiences while researching my books – PEH.
[4] H.B. Cott, *Looking at Animals: A Zoologist in Africa*, Collins, London, 1975.
[5] See N. Tinbergen, *Curious Naturalists*, Penguin, 1974.
[6] *Ibid*.
[7] Internet site: "Diogenes", 2010 http://hubpages.com/hub/LONOMIA-The-Killer-Caterpillar.[8] J.C. Miller, D.H. Janzen and W. Hallwachs, *100 Butterflies and Moths: Portraits from the Tropical Forests of Costa Rica*, Belknap Press, Harvard, 2007.

Chapter 3

[1] Paul-Henri Fabre was born in 1888, so it is likely that "*La Soirée du Grand-paon*" referred to here was in the mid 1890s.
[2] The chemist, A. Butenandt, who later identified the first sex pheromone from the female common silkworm received the Nobel Prize in 1939 for his work on the chemistry of hormones.
[3] The term "pheromone" is used to describe substances used in communication between organisms of the same species, acting externally, in contrast to hormones, which act internally.
[4] M.E. Tolliver & M.R. Jeffords, "Long distance dispersal by *Callosamia promethea* (Saturniidae)", *J. Lepid. Soc.*, 1981, 35:76.
[5] W.J.B. Crotch, *A Silkmoth Rearer's Handbook*, 1956, Amateur Entomologists' Society, UK.
[6] J.C. Regier et al., "Phylogenetic relationships of wild silkmoths (Lepidoptera: Saturniidae) inferred from four protein-coding nuclear genes", *Systematic Entomology*, 2008, p.1365.
[7] J.S. McElfresh & J.G. Millar. *J. Chem. Ecol.*, 25, 2005.

Chapter 4

[1] Ovid, *Metamorphoses*.
[2] Conspicuousness, not eye mimicry, makes "eye-spots" effective antipredator signals. M. Stevens, C.J. Hardman and C.L. Stubbins, *Behavioural Ecology*, 2008.
[3] Altmann, S.A., "Avian mobbing behavior and predator recognition", *The Condor*, 1956, 58:241.
[4] McPherson & Brown, *Raptor* 15, Dec. 15 1981.
[5] J.J. Negro, G.R. Bortolotti and J.H Sarasola, "Deceptive plumage signals in birds: manipulation of predators or prey?", Internet Publication.
[6] Blest had not noticed that the small tortoiseshell (*Aglais urticae*) has a white-black-yellow banding pattern on its wings that mimics the colours of the bodies of several common European bumble-bees.
[7] A.D. Blest, "The function of eyespot patterns in the Lepidoptera", *Behaviour 11*, 1957, p.209.
[8] M. Olofsson, A. Vallin, S. Jakobsson, C. Wiklund, "Marginal Eyespots on Butterfly Wings Deflect Bird Attacks Under Low Light Intensities with UV Wavelengths", *PLoS ONE* 5(5): e10798, Internet Publication.
[9] David Stradling first suggested that they were frog mimics. I argued in *Butterflies: Messages from Psyche* that they are toad mimics, unaware at the time of Stradling's paper: D.J. Stradling, "The Nature of the Mimetic Patterns of the Brassolid Genera, *Caligo* and *Eryphanis*", *Ecological Entomology* Vol. 1, 135.
[10] "I have seen bats take large numbers of saturniids right before my eyes. Bats don't pay any attention to eye-spots, and they can be deadly for moths. One night my wife and I were mothing with a light far from civilisation in a desert wash in Baja California. We were searching for the rare *Agapema pelora* moth of which only seven male specimens were known. Soon after dark the males began to arrive along with the bats, which snatched several right before my eyes before I began waving my butterfly net at them. They were so close that one actually got caught in my net. They continued to eat incoming males in the distance, but several moths got through to land on the white sheet where they could be captured. Then my wife called me for dinner, which I took back to the light to watch as I ate. For the ten minutes it took me to wolf down my dinner, the bats got all incomers. I quickly resumed chasing the bats and caught several more male moths when suddenly the coveted female we sought plunked down on the sheet. She gave us many eggs which we reared to adults and so we were able to film all stages of development and share these rare specimens with museums around the world. But the battle of the bats demonstrated to me the tremendous impact they have on nocturnal moths." Kirby Wolfe.
[11] J.E. Yack, "Seasonal partitioning of atympanate moths in relation to bat activity", *Canadian Journal of Zoology*, 1988, 66:753-755.
[12] Scott B. Morrill, James H. Fullard, "Auditory influences on the flight behaviour of moths in a Nearctic site", I. Flight tendency, *Canadian Journal of Zoology*, 1992, 70:1097.
[13] J.H. Fullard & N. Napoleone, "Diel flight periodicity and the evolution of auditory defences in the Macrolepidoptera", *Animal Behaviour*, 2001, 62, 349.
[14] There is a rare variety of the emperor moth (*Saturnia pavonia*) that lacks eye-spots.

Chapter 5

[1] A.D. Blest, "The Evolution of Protective Displays in the Saturnioidae and Sphingidae (Lepidoptera)", *Behaviour 11*, 1957, pp. 257-309.
[2] A.D. Blest, "The Function of Eyespot Patterns in the Lepidoptera", *Behaviour 11*, 1957, 209.

Chapter 6

[1] Isabel is the Spanish spelling.
[2] See Additional Notes.
[3] J. Regier et al., "Phylogenetic relationships of wild silkmoths. (Lepidoptera: Saturniidae) inferred from four protein-coding nuclear genes", *Systematic Entomology*, 2008, 1365-3113.
[4] In W.J.B. Crotch, *A Silkmoth Rearer's Handbook*, Amateur Entomologists' Soc., 1956.

Chapter 7

[1] Naomi Pierce, co-author "Phylogeny and Palaeoecology of *Polyommatus* blue butterflies, *Proceedings of the Royal Society*, B, Feb. 2011.
[2] D. Janzen, "Weather-Related Color Polymorphism of *Rothschildia lebeau* (Saturniidae)", *Bulletin of the Entomological Society of America*, Vol. 30, 1984.
[3] F. Hernández-Chavarría, A. Hernández & A. Sittenfeld, "The 'windows', scales, and bristles of the tropical moth *Rothschildia lebeau* (Lepidoptera:Saturniidae)", *Revista de Biología Tropical*, Costa Rica, 2007.
[4] C. Beutelspacher, *Mariposas entre los Antiguos Mexicanos*, Tezontle, UNAM Mexico, 1989.
[5] The ground colour of Rothschildias varies between the wet and dry season in some species, as shown by Janzen (*loc. cit.*) for *R. Lebeau*.
[6] Donald A. Mackenzie, *Precolumbian America: Myths and Legends*, Senate, 1995.
[7] It is very tempting to make a connection here with the book of Gavin Menzies (*1421: The Year the Chinese discovered the World*, Bantam, 2002). He found convincing evidence for Chinese treasure fleets having traversed the globe in 1421-3 about a century before the arrival of Cortés. A number of the giant junks came to grief on the coasts of the New World, and there is evidence for survivors establishing settlements on the Pacific and Atlantic seaboards and being incorporated into local communities along with hens, pigs, etc. that were rescued from the shipwrecks.
[8] According to E. Velikovsky's *Worlds in Collision* the plumed serpent and Chinese dragon and Jewish golden bull represent a wayward Venus which threatened earth on many occasions before colliding with Mars, thus acquiring its present orbit.
[9] D.L. Evans, "Defensive behavior in *Callosamia promethea* and *Hyalophora cecropia* (Lepidoptera: Saturniidae)", *American Midland Naturalist* 100, (1978).

Chapter 8

[1] Philippa Scott, *Silk*, Thames & Hudson, 1993.
[2] This last is used for silk by Native Americans in Mexico and Guatemala and other New World peoples. These are generally known as "tent caterpillars".
[3] This is how it was sometimes seen in mediaeval times in Europe.
[4] See Richard Peigler, "Moth Cocoon Artifacts", *Cultural Entomology Digest* 4 (Internet source).
[5] P.E. Howse, *Butterflies: Messages from Psyche/Decoding their Signs and Symbols*, Papadakis 2010.
[6] According to R. Peigler this was silk made from members of the Lasciocampidae.
[7] C.R. Beutelspacher, *Las Mariposas entre los Antiguos Mexicanos*, UNAM Mexico, 1989.
[8] Philippa Scott, *Ibid*.
[9] Eva Panagiotakopulu, "Butterflies, Flowers and Aegean Iconography: A Story about Silk and Cotton, www.therafoundation.org.
[10] G. Menzies, *1421. The year China discovered the World*, Bantam Press, London, 2002.

FURTHER READING

Baxter, R.N. *Rearing Wild Silkmoths*. Chudleigh Publishing, Ilford 1992

Collins, M.M. & R.D. Weast, *Wild Silk Moths of the United States*. Collins Radio Co., Cedar Rapids 1961

Feltwell, J. *The Story of Silk*. Alan Sutton, Gloucester 1990

Forbes, P. *Dazzled and Deceived*. Yale University Press, New Haven and London 2009

Gardiner, B.G.C. *A Silkmoth Rearer's Handbook*. The Amateur Entomologists' Society, Hanworth 1982

Howse, P.E. *Butterflies: Messages from Psyche*. Papadakis, London 2010 (also published as *Butterflies: Decoding their Signs & Symbols*. Firefly, New York and Ontario 2010)

Miller, J.J., D.H. Janzen & W. Hallwachs (eds.), *100 Butterflies and Moths. Portraits from the Tropical Forests of Costa Rica*. Belknap Press of Harvard University Press, Cambridge, Massachusetts 2007

Peigler, R. & J. Cody, *Wings of Paradise: The Great Saturniid Moths*. University of North Carolina Press, Chapel Hill 1996

Pinhey, E. *Emperor Moths of South and Central Africa*. C. Struik (Pty) Ltd, Cape Town 1972

Scott, Philippa, *The Book of Silk*. Thames and Hudson, London 1993

Tuskes, P.M., J.F. Tuttle & M.M. Collins, *The Wild Silk Moths of North America: A Natural History of the Saturniidae of the United States and Canada*. Cornell University Press, New York 1996

Wolfe, K.L. *The Saturniidae: Wild Silkmoths/Giant Silkmoths/Emperor Moths*. Self Publ. Escondido, California 2005, CD

ADDITIONAL NOTES

On biology, classification, rearing and photographing silkmoths

Kirby Wolfe

Definitions Moths and butterflies belong to the group of insects called Lepidoptera Although we often think of moths as nocturnal, or flying at night while butterflies are diurnal, flying in the day time, some moths are also diurnal. And even though some moths are small and dull-coloured, others are brightly coloured or large like butterflies. So what is the difference between moths and butterflies? A general rule is that the tip of the butterfly's antenna is the thickest part whereas the tip of the moth's antenna is the most slender.

The Lepidoptera are divided into families, one of which is the Saturniidae. In turn, the family Saturniidae is divided into subfamilies, and further into tribes, genera, sometimes subgenera, species and subspecies. This system of organization will be further treated below.

It is not known how long ago the Saturniidae may have branched off from other families, but fossil evidence suggests that the Lepidoptera in general blossomed during the Cretaceous period, which ended 65 million years ago. Today, the Saturniidae can be recognised by their antennae, which are generally rather flat and feathery or pectinate, which means "comb-like". Most moths of this family have either two or four pectens or "teeth" on each antennal segment, and females have shorter ones.

Reproduction Saturniids do not live long as adults, usually from several days to about one week, as they do not have functional mouthparts in most cases and so cannot feed on flowers as do hawk moths and butterflies. The main function of adult saturniids is to mate and distribute their eggs. The female first emerges from her pupa, which is the hard casing inside a cocoon or buried in the soil or in loose debris. She then pumps body fluid into her wing veins to hydraulically extend them. When they are dry she extends a small spike at the tip of her abdomen, or bulbous third section of her body. From the spike she emits a tiny gaseous plume of chemicals called pheromone, which floats downwind. This is done during a particular time of day or night, depending on the species, and often lasts only one or two hours. Males, which have emerged from their cocoons, begin searching for females during the prescribed time for each species. Mating may last for only a few minutes or many hours, depending on the species. Then the female flies off to deposit the eggs, usually in groups on the leaves or stems of particular plants, as most species of saturniids are specialized in their choice of food plant.

Life cycle When the caterpillars (larvae) "eclose" or hatch from their eggs, they begin to feed, often in tight groups, taking turns to eat at the freshly chewed leaf edge. This first stage, or instar, usually lasts about a week as the larva fills its first "skin". Then it stops feeding for a day or so when the old "skin" splits open and the second instar larva crawls out. It may or may not resemble the first instar. This process is repeated usually through 4-7 instars, depending on the species. The larva then spins a silk cocoon, or crawls away to dig into the ground or debris. All species form a hard pupa where they await the proper season and weather conditions to emerge and reproduce. The pupal stage may be as short as several weeks. Tropical species may have several broods per year, but in species from the far north or far south, where winters are cold and vegetation scarce at that time, a pupa may remain dormant for a year or more. In desert regions, where rainfall is unpredictable, pupal emergence may be staggered over many years, increasing the probability that some adult moths may emerge during a year in which the caterpillars will have something to eat. I once reared a group of caterpillars of Burns' buckmoth (*Hemileuca burnsi*), and their pupae just lay there, with several adults emerging every year for nine years!

Dangers for moths I am often asked what is the purpose of these moths if they live only a few days as adults. That is a philosophical question, but as I have observed with life in general, the outcome is to be eaten at one stage or another. In the Saturniidae, all stages are targeted. A female moth may produce 100 to 200 eggs or more, but few reach adulthood. Eggs are attacked by tiny wasps that may deposit their own eggs in some or all of the eggs in a group, and their offspring feed off the egg contents. Caterpillars are also parasitized by several other types of wasps, some tiny and some large, or are eaten by even larger wasps or other insect predators, birds and mice. The pupa is not immune from these dangers. The most common dangers for adult moths are bats, birds and electric lights, to which many are attracted and marooned until daylight, when birds are likely to find them. Humans, through deforestation, introduction of exotic parasites and widespread use of insecticides, are even greater threats to these mainly inoffensive beautiful silkmoths.

Eating saturniids Caterpillars of several species of saturniids are regularly eaten. Previously,

Native Americans ate the caterpillars of the pine moth, *Coloradia pandora*, migrating to the mountains of the western United States to gather the abundant larvae, which they dried and put into their soups as flavouring. In Africa various species are eaten, and are commonly sold in some markets. The most famous African caterpillar eaten is the "mopane worm", from the mopane moth, *Gonimbrasia belina*. These also are mostly dried and used in soups.

Irritating and poisonous saturniids
Although the vast majority of the more than 1500 saturniid species are inoffensive and provide food for birds, bats and other life forms, a handful are pests of one form or another. Several American tropical species are directly harmful to humans. Females of most *Hylesia* species, a genus of mostly tiny, rather drab saturniids, have their abdomen armed with stiffened and sharpened hairs from which they form a matting to cover their egg masses in order to discourage predation. When the moths flutter, as around porch lights, some of these tiny "lancets" are shed, and can be very irritating to humans, causing severe dermatitis in some individuals. In areas of Brazil and other countries where these moths have a season of great abundance, lights may be left off or bedclothes folded and put away until bedtime, to avoid contact with the moths and their irritating hairs.

Agricultural pests Surprisingly few saturniid caterpillars are important agricultural pests. In the western United States, at least three species are occasionally problematic. The most important, the range caterpillar, *Hemileuca oliviae*, which feeds upon grasses on grazing lands, can significantly reduce available feed for cattle. Caterpillars can reach densities of 100-300 per square metre (yard), stinging cattle and contributing to soil erosion. The pine moth, *Coloradia pandora*, mentioned above as a food source for Native Americans, periodically causes extensive defoliation of pine trees throughout the West. Caterpillars of another member of this subfamily, *Hemileuca eglanterina*, sometimes cause damage in fruit orchards in California's central valley. In Chile, the beautiful *Polythysana rubrescens* has become a pest of the farmed forests of non-native Monterey pine trees planted along the central coast. And in Africa, the Pine Emperor, *Imbrasia cytheraea*, has sometimes become a pest of several kinds of non-native pines.

COLLECTING MOTHS

Collecting has two common meanings when applied to insects: to capture or to make a collection of. There are various reasons for collecting silkmoths. Scientists, hobbyists, students, artists, including photographers, are some of the types of people who may collect these beautiful insects. Collecting moths and other insects has often provided early stimulation for an appreciation of nature and wildlife, and many children who first collected insects have gone on to become natural scientists as adults. Elementary school teachers like to demonstrate the metamorphosis of caterpillar to moth in the classroom, to the wonderment of young students.

For serious collecting, light traps emitting high intensity light extending into the near ultraviolet can be obtained from biological suppliers. Methods of killing, mounting and labelling moths can be found in good insect identification guides, but most naturalists get great satisfaction from photographing the live insects (see below).

REARING MOTHS

Rearing saturniid moths is the best way to obtain perfect specimens for photography or a collection. If they are reared then killed after drying their wings but before flying, the resulting specimens are much more colourful and beautiful to contemplate. In most cases, rearing is not difficult, and one can really enjoy watching the caterpillars eat and grow, many species changing colours or spines with each successive instar or stage. And the satisfaction of success and new knowledge is an additional reward.

Equipment and needs Caterpillar rearing can be as high or low tech as you wish, provided that a few major rules are followed. First and foremost, do not try to rear too many at once. This is a temptation with beginners, as it can seem so easy at first, but in overcrowded conditions disease will usually flourish before the caterpillars reach maturity, which leads to discouragement. I try to keep numbers between 12 and 20 caterpillars per container.

For native species, housing the growing caterpillars can be as easy as placing them in a large sleeve of open-weave fabric or screen attached to a branch or shrub outdoors. There are some dangers with this method, including predation by birds, ants, spiders and wasps, and people, to name a few. Non-native or exotic species should never be reared outdoors, where they might escape.

All species can be reared indoors, in plastic boxes of various sizes with tight-fitting lids. Rectangular openings can be cut into the lids and appropriate pieces of plastic screen of the several available mesh sizes can be attached with a glue gun. For better viewing, an aquarium can be fitted with a lid and partly covered with glass to maintain moisture, as indoor conditions are usually much drier than outdoors. Additionally, potted plants or small trees can be temporarily brought indoors and sleeved.

Eggs Wild caterpillars can sometimes be found, but beginners may not be able to distinguish between saturniid caterpillars and other types. But if an egg-laden female saturniid is discovered, fortunately it is usually easy to obtain eggs, because she deposits them all within a few days. The surest way is to place her in a small paper bag and let her deposit the eggs. She will damage herself in a bag, so if she is desired as a specimen, she can be placed with wings folded in a glassine envelope. This way she will usually lay without damaging herself.

Once the eggs are gathered, I place them in a plastic covered petri dish alongside a small wad of wet paper towel to maintain moisture, because most eggs will dry out and die indoors without added moisture. Most types of eggs hatch within one to three weeks, depending on the species, although some require much longer, or even a period of refrigeration. Eggs of a species of *Hylesia* I reared hatched after thirteen months under normal conditions.

When the eggs hatch, I remove the tiny larvae with a small, soft camel hair pointed brush and transfer them to a small plastic box with its expected foodplant in a plastic corsage tube filled with water. Any small container can be used, but you must seal all around the plant stem with something like pieces of paper towel or the caterpillars will stupidly walk down into the water and drown.

Caterpillar food While we wait for the eggs to hatch, we should think about what to feed the babies. Fortunately, there are several books available to help, including Steven Stone's *Foodplants of world Saturniidae* and Brian Gardner's *A Silkmoth Rearer's Handbook*. A person who becomes an avid caterpillar rearer has to become familiar with trees and other possible food plants. Most saturniid species specialise on one or a very few plants, but many will accept a few alternatives, especially from the same plant genus or sometimes the same plant family or even other families. I keep a tiny nursery with a variety of potted plants, and when I have caterpillars whose natural hostplant is unknown, I make a salad when they hatch, cutting the tips of some leaves to draw fresh, inviting juice to tempt them to feed. Sometimes they refuse everything, a sad day.

As the larvae grow larger, they have to be moved to larger boxes. I place a paper towel in the bottom and fashion a larger water container. Eventually the larvae will be eating prodigious amounts, and must not be allowed to run out of food even for a few hours, or they will result in undersized adults. Fortunately, this last phase rarely lasts longer than one to two weeks and the whole time from hatching egg to cessation of feeding is generally three to six weeks.

Pupation When caterpillars are large and fat and have finished feeding they will often rest motionless for a period, changing colour or shape, then will look for a place to spin a cocoon or, in the case of species that do not spin cocoons, will wander aimlessly before hiding in an appropriate place to become a pupa. Among this group are those that burrow into the ground and those that pupate among leaf litter bound together loosely by silk threads. The cocoon makers can be left alone for about ten days, by which time they have become hardened pupae inside, when they may be gathered and stored. The burrowers and the species with bare pupae from wet tropical climates I place in small plastic boxes with several layers of dampened paper towel cut to size, inside a closed cardboard box in a dark closet. Only caterpillars of the same stage on the same day should be placed together, up to three or four per box. They should pupate in about nine days.

With pupae, there are still important requirements for keeping them alive and healthy until the time for adult emergence. All species may be kept in aquariums, but with different humidity levels for species from arid and humid climates, and always out of direct sunlight. I use 40-litre (10-gallon) aquariums with about 25 mm (1 in.) vermiculite planting material, which I keep more or less damp. A piece of fiberglass window screen is draped down inside the front glass, across the bottom and up the rear glass and fixed at the top and sides with a glue gun. On this I place small plastic boxes mostly filled with vermiculite more or less humid, upon which I lay the bare pupae or stack cocoons opening end up or on their sides. Long-tailed or very large moths are best hung on the screen at the back or by other means. The aquariums are covered by a framed screen with a pane of glass on top to adjust the amount of airflow, to control humidity. I use distilled water to maintain humidity to avoid salt build-up, and spray water on all cocoons and pupae about twice a week.

Waiting for cocoons and pupae to produce adult moths sometimes requires patience. Although under certain circumstances many species will emerge within three to six weeks, others emerge only annually, if that. In the western United States, the ceanothus silkmoth often remains in the cocoon for two years. Some desert species can take much longer (Burn's buckmoth, above).

In the dry western deserts of California, some species spread out their emergence times over a number of years in order to increase the possibility that sufficient rain will have fallen during at least one of those years to produce leaves on the caterpillar's hostplant. In the case of the Burns' buckmoth, *Hemileuca burnsi*, I once reared several dozen pupae of this attractive moth, and although a few adults emerged the following and successive years during its normal season, I had to wait nine years until the last one emerged.

Slumbering pupae are said to be in diapause. Many tropical species do not diapause, and emerge as soon as the caterpillar can convert to a moth inside the pupa, while others do diapause and require specific stimulation to "break diapause". The precise stimulation varies between species, often seemingly consisting of a combination of factors, largely unknown. Photoperiod, or daylight length, is one known; atmospheric pressure is another. Rearing moths indoors where electric lights artificially lengthen the day causes some moth species that normally diapause until longer daylight hours to be fooled and emerge much sooner than normal. For tropical species, a sudden lower air pressure mimics a tropical storm, which can cause them to emerge indoors when it rains outdoors.

Emergence Moths emerge at different times of the day or night, depending on species. Ideally, we would like to have the wings well hardened but before any flight has occurred when we attempt to photograph or kill the specimen. This is not always easy to accomplish with moths that emerge in the middle of the night. Sometimes they may be ruined by morning and other times we get lucky. For moths that emerge during the day but wait to fly at night, we have a chance to move them to a lighted cage before dark, where they will usually remain motionless sometimes for several days or more.

PHOTOGRAPHING MOTHS

Silkmoths are generally easier to photograph than butterflies and many other live subjects, because most are nocturnal and are reluctant to fly during the day unless placed in sunshine. They can usually be positioned on a pleasing or natural background and will remain there while being photographed, especially if kept out of direct sunlight.

The first objective is to obtain a perfect specimen, because scratches or worn imperfections will become very obvious and can spoil an otherwise perfect photograph. A small, smooth but slightly roughened round stick smaller than a pencil can be used to transport the moth. I use the handle end of a small artists' brush, with the paint removed from the tip and the wood barely roughened to provide a foothold. This is placed behind and parallel to the body and moved forward between the moth's legs, which are teased one by one with the tip until most or all legs are grasping the handle. The moth hangs from the handle and can be transferred to a cage or to a photographic perch or background where the process is reversed until all legs are grasping the desired object. The moth can be placed under a light until proper conditions for a photograph are available. Difficult or day-flying moths may be placed in the refrigerator to await photographing.

Let's consider the camera equipment. Although it is possible to produce pleasing photographs with any number of cameras, there are several requirements for achieving consistently good photographs. Unfortunately, this requires money. With cameras, you get what you pay for.

For best results, a single lens reflex camera, either film or digital, is almost a requirement, allowing the photographer to see the subject through the lens, not only to assure sharp focus but also to allow different lenses to be used. For macro, meaning close-up work, focus is critical, and for this manual focus is preferable to automatic focus, even with an auto focus camera. Conversely, automatic exposure is very helpful for consistent proper exposure, to avoid a photograph that is either too dark or too light. A macro lens, which generally costs slightly more than a standard lens, is a prerequisite for best results. This allows you to focus more closely in order to fill the frame with your subject. An electronic flash is often useful if carefully placed and controlled. Built-in flashes are not useful for macro.

If you do not ever want to make large prints, relatively inexpensive consumer models of both film and digital cameras will suffice for many conditions. Each year the capability of digital improves, and who knows what the future will bring.

For film cameras, most amateurs use 35 mm which uses a normal 50 mm lens. Although 50 mm macro lenses are adequate, the 100 mm macro lens is better, as it allows the camera to be further from the subject, allowing more light to reach the subject. Because many digital cameras presently have a sensor smaller than 35 mm film, a 50 mm macro lens will be more powerful on those models and can probably be used as a compromise.

Most macro photographs were formerly made entirely with flash, which often resulted in a black background. Black backgrounds are now giving way to lighter, natural looking backgrounds since modern technology made this easier for everyone. But first, for those unfamiliar with camera jargon, let's quickly review some of the basics.

The camera is like a box with film or a digital sensor in the back and in front a lens that allows light to enter and focuses it on the film or sensor. Between the front of the lens and the film are two separate systems that control the light. The **aperture** is the adjustable size of the opening in the lens through which the light passes; the larger the aperture, the more light enters. Confusingly, the aperture settings are represented by what are known as f stop numbers, which close the aperture increasingly the higher the number. A camera that has a lens of f2.8 to f32 means that the aperture is wide open at f2.8 but is closed as much as possible at f32. At f11 it is partly closed. But the light also has to pass through the **shutter**, which opens and closes at different speeds, usually measured in fractions of a second. Photographers traditionally had to calculate aperture and shutter speed using light meters, but with automatic exposure the results are much more reliable.

One of the things that remain to be decided by the photographer is the depth of field, or depth of focus. Sometimes it is best to have an out-of-focus background, thereby highlighting the subject. Other times, maximum depth of field is of most importance. This is especially true in macro photography, where depth of field is much shallower. Because of the way a lens is shaped, the smaller the aperture the greater the depth of field, or area in sharp focus. I like to use f22 aperture for most macro photos, but that often results in long exposure times. Therefore a sturdy tripod, and I emphasise sturdy, is a necessity for this type of photography, along with a cable shutter release to avoid movement of the camera.

To achieve great depth of field and natural backgrounds, try to obtain a camera with **aperture priority**, which means that you decide the aperture and the camera calculates the shutter speed based on the light coming from the centre of the photograph. Next is lighting. You will find that macro photography in full sunlight produces extreme bright and shadow areas, and without flash to help overcome this, you must avoid direct sunlight. There are several alternatives. The first, shooting in open shade, bathes everything with blue from the sky, and is not satisfactory. Cloudy days avoid most of the blue colour, but can be somewhat lacking in contrasts and vibrant colours. As a compromise, you can shoot under a white sheet or umbrella in open sun, trying to reflect sunlight onto the subject under the sheet with white posterboard or gauze-covered mirrors. Using a digital camera can provide instant feedback for perfecting these techniques.

I use several tripods and improvised clamps for holding subjects and reflectors, and sometimes electronic flash. When all is in place, I bring the moth to its place under the sheet and focus on it. At that point, the moth is usually resting with closed wings. Male moths will often open their wings if you blow on them, more or less gently. Females are more difficult, but their wings can often be raised with a finger. If they begin vibrating, they are warming their engines for flight, and will usually continue until suddenly going airborne. Blowing can sometimes stop them for a few seconds, but usually you will have to quit at that point and try again later.

Everyone has a different reason for taking photographs. The final result you desire will dictate what minimum equipment you require. Photography magazines and books are available for further information.

CLASSIFICATION

Saturniid evolution and classification

Taxonomy is the science of classification. It attempts to describe, identify or name organisms and arrange them into hierarchical groups according to their characteristics and relationships. Anyone who has taken a high school biology course has had to learn the basic traditional system. For the Saturniidae, we use several additional divisions. This would be the classification for the **American luna moth:**

Kingdom:	Animalia	
Phylum:	Arthropoda	
Class:	Insecta	
Order:	Lepidoptera	
Family:	Saturniidae	(Sa-tur-nye'-i-day)
Subfamily:	Saturniinae	(Sa-tur-nye'-i-nay)
Tribe:	Saturniini	(Sa-tur-nye'-i-nye)
Genus:	*Actias*	(always capitalised)
Subgenus:	------	
Species:	*luna*	(not capitalised)

This arrangement is sometimes shown as a family tree, which suggests that different species are related, meaning that at some point in the past they shared an ancestor; a fork of a branch, so to speak. And if we follow the branch further back, we come to further forks, and so on until we reach the main trunk. All the tips of the branches represent separate species, but they all share the same trunk. They are made of roughly the same DNA.

Many lepidopterists have used moth anatomy, shape and wing design, to base theories about where and when the species were formed, which are most primitive, more like the ancestor, and which are more recently developed. They study the present distribution of the moths and postulate how and when the different species may have arrived where they are today. Other clues to compare are the caterpillars, which in many cases are similar to their closest relatives even if they live far away.

For some closely related or similar species, dissection of the genitalia can often reveal the most obvious clues to distinguish them, and all taxonomists and many serious students learn how to do this fairly simple operation.

More recently scientists have used nuclear gene sequencing to graphically illustrate genetic differences and similarities between species, thereby demonstrating their relationships with more certainty than ever.

Now, one of the main obstacles to taxonomy has become a misguided effort, based on the excuse of protecting endangered species, to prohibit or severely limit insect collecting and study in many tropical countries where we know the least about the Lepidoptera fauna. If habitat is not destroyed, there is no danger to insects from collecting. If habitat is destroyed, no amount of legal protection will prevent the disappearance of most insect species in those areas.

Origin and subfamilies South America possesses five major subfamilies of saturniids and the rest of the world has only one major and two minor subfamilies. From this original saturniid trunk we now have five subfamilies (**Oxyteninae**, **Cercophaninae**, **Arsenurinae**, **Ceratocampinae** and **Hemileucinae**) that seem to have developed only on the South American continent. These five subfamilies were then able to extend to North America only after the Central American land bridge was in place. At some point, a tiny branch called the **Agliinae**, with only one genus, spread to Eurasia. Another, the **Salassinae,** also with only one genus, developed in southern Asia. The most important branching, the **Saturniinae** occurred in either Africa or Eurasia; it includes most of the well-known giant silkmoths. One branchlet became the tailed moon moths, with smaller twigs represented by the African and Madagascan moon moths (*Argema*), a wide variety of south Asian species of *Actias* moon moths, and America's *Actias* luna moth. The Spanish moon moth, *Graellsia isabelae*, developed in the pine forests of Spain, France and Italy. From this point there are branches in Africa or Eurasia to all the rest of the wild silkmoths that populate Australia, Asia, Europe, Japan, Africa, and many islands.

Tribes The subfamily Saturniinae is divided into tribes, with the African tribe **Bunaeini** being perhaps the most primitive along with fellow Africans **Micragonini** and **Ludiini**. The remaining branch forks widely between the remaining tribes **Saturniini** and **Attacini**, some of which are the really giant silkmoths that often possess angular translucent patches on their wings and are usually coloured in combinations of reddish, black and white. The saturniini genera are more likely to have the typical saturnine rings on all wings. One saturniini genus that causes much speculation is *Copaxa*, my specialty, whose ancestry and distribution is a mystery. It is the only endemic genus of the saturniini tribe on the American continents, and probably developed in South America. Perhaps its ancestors may have been blown to South America from North Africa or Laurasia when the continents were closer together. Nuclear sequencing by Regier, et al. suggests that the South American species *Rothschildia* is closely related to its similar relatives *Archaeoattacus* and *Attacus*. So the ancestor of *Rothschildia* may also have come from Africa or Laurasia. But all of this is speculation.

The following chart shows the classification down to genus level of the saturniids presented in this book, based on clues of their proximity to an ancestral form. Subfamilies and tribes are in bold.

- **Cercophaninae**
  - Cercophana
  - Janiodes
- **Arsenurinae**
  - Arsenura
  - Caio
  - Dysdaemonia
  - Titaea
  - Paradaemonia
  - Rhescyntis
  - Copiopteryx
  - Loxolomia
- **Ceratocampinae**
  - Eacles
  - Bathyphlebia
  - Citheronia
  - Citheronula
  - Othorene
  - Psilopygida
    - Psigida (subgenus)
  - Syssphinx
  - Anisota
  - Adeloneivaia
  - Adelowalkeria
  - Citioica
  - Ptiloscola
- **Hemileucinae**
  - **Hemileucini**
    - Lonomia
    - Periga
    - Coloradia
    - Hemileuca
    - Hylesia
    - Gamelia
    - Hyperchiria
    - Automeris
    - Leucanella
    - Pseudautomeris
    - Molippa
    - Ormiscodes
    - Paradirphia
    - Cerodirphia
    - Dirphia
    - Periphoba
    - Dirphiopsis
    - Pseudodirphia
    - Adetomeris
  - **Polythysanini**
    - Polythysana
- **Agliinae**
  - Aglia
- **Salassinae**
  - Salassa
- **Saturniinae**
  - **Micragonini**
    - Micragone
    - Holocerina
    - Vegetia
    - Goodia
  - **Bunaeini**
    - Eochroa
    - Melanocera
    - Cinabra
    - Pseudobunaea
    - Cirina
    - Bunaea
    - Gynanisa
    - Gonimbrasia
    - Imbrasia
    - Lobobunaea
    - Aurivillius
    - Bunaeopsis
  - **Saturniini**
    - Antherina
    - Antheraea
    - Opodiphthera
    - Actias
    - Graellsia
    - Argema
    - Loepat
    - Cricula
    - Copaxa
    - Saturnia
      - Neoris (subgenus)
      - Caligula (subgenus)
      - Perisoena (subgenus)
      - Saturnia (subgenus)
      - Eudia (subgenus)
      - Calosaturnia (subgenus)
      - Agapema (subgenus)
  - **Attacini**
    - Rhodinia
    - Rothschildia
    - Eupackardia
    - Callosamia
    - Hyalophora
    - Samia
    - Archaeoattacus
    - Attacus
    - Coscinocera
    - Epiphora

INDEX